The Kubernetes Operator Framework Book

Overcome complex Kubernetes cluster management challenges with automation toolkits

Michael Dame

BIRMINGHAM—MUMBAI

The Kubernetes Operator Framework Book

Copyright © 2022 Packt Publishing

Group Product Manager: Rahul Nair
Publishing Product Manager: Surbhi Suman
Senior Editor: Sapuni Rishana Athiko
Content Development Editor: Yasir Ali Khan
Technical Editor: Shruthi Shetty
Copy Editor: Safis Editing
Project Coordinator: Shagun Saini
Proofreader: Safis Editing
Indexer: Rekha Nair
Production Designer: Prashant Ghare
Marketing Coordinator: Nimisha Dua
Senior Marketing Coordinator: Sanjana Gupta

First published: June 2022
Production reference: 1070622

Published by Packt Publishing Ltd.
Livery Place
35 Livery Street
Birmingham
B3 2PB, UK.

ISBN 978-1-80323-285-0

www.packt.com

To my wife, Meaghan, my parents and sister, the boys of Gold Team and Gamma Tetarton, and all of the amazing mentors and colleagues who have guided me along in my career, helping me to grow into a better engineer.

– Mike Dame

Contributors

About the author

Michael Dame is a software engineer who has several years of experience in working with Kubernetes and Operators. He is an active contributor to the Kubernetes project, focusing his work on the Scheduler and Descheduler sub-projects, for which he received a 2021 Kubernetes Contributor Award. Michael has a degree in computer science from Rensselaer Polytechnic Institute and has worked on the Kubernetes control plane and OpenShift Operators as a senior engineer at Red Hat. He currently works at Google as a software engineer for Google Cloud Platform.

I want to thank Meaghan, for supporting and encouraging me throughout the writing process, and the many brilliant colleagues who took the time and had the patience to teach me everything I know about Kubernetes.

About the reviewers

Mustafa Musaji is an experienced architect and engineer in the enterprise IT space. He develops and designs open source software solutions for large organizations looking to adopt transformational changes within their IT estate. Mustafa currently works at Red Hat as an associate principal solutions architect, where he supports customers in the journey to cloud-native applications. That involves not only discussing using container-based deployments but also the underlying orchestration and dependent services.

Drew Hodun is the co-author of O'Reilly's *Google Cloud Cookbook*. He has had a number of roles at Google, both customer-facing and in product engineering. His focus has been on machine learning applications running on the Google cloud and Kubernetes for various clients, including in the autonomous vehicle, financial services, and media and entertainment industries. Currently, Drew works as a machine learning tech lead working on AI-powered EdTech products. He has spoken at a variety of conferences, including Predictive Analytics World, Google Cloud Next, and O'Reilly's AI Conference. (Twitter: @DrewHodun)

Thomas Lloyd III is an information technology professional who specializes in systems and network administration. He currently works in the non-profit sector, lending his skills and experience to St. Catherine's Center for Children located in Albany, NY. Prior to this. he worked for a variety of private sector companies, where he honed his skill set. He has an A.A.S. in systems and network administration and is currently in the process of earning his B.S. in technology from SUNY Empire.

I'd like to thank the countless individuals who have helped me throughout my education and career. I wouldn't be where I am today without the endless support of my family, friends, professors, and many colleagues who gave me guidance and the opportunity to succeed.

Table of Contents

Preface

Part 1: Essentials of Operators and the Operator Framework

1

Introducing the Operator Framework

Technical requirements	4	Managing Operators with OLM	11
Managing clusters without Operators	4	Distributing Operators on OperatorHub.io	12
Demonstrating on a sample application	4	Defining Operator functions with the Capability Model	14
Reacting to changing cluster states	5		
Introducing the Operator Framework	6	Using Operators to manage applications	19
Exploring Kubernetes controllers	6	Summarizing the Operator Framework	19
Knowing key terms for Operators	7	Applying Operator capabilities	20
Putting it all together	8	Summary	22
Developing with the Operator SDK	9		

2
Understanding How Operators Interact with Kubernetes

Interacting with Kubernetes cluster resources 26
Pods, ReplicaSets, and Deployments 27
Custom resource definitions 28
ServiceAccounts, roles, and RoleBindings (RBAC) 29
Namespaces 31

Identifying users and maintainers 32
Cluster administrators 32
Cluster users 33

End users and customers 35
Maintainers 36

Designing beneficial features for your operator 36
Planning for changes in your Operator 38
Starting small 38
Iterating effectively 39
Deprecating gracefully 40

Summary 40

Part 2: Designing and Developing an Operator

3
Designing an Operator – CRD, API, and Target Reconciliation

Describing the problem 46
Designing an API and a CRD 48
Following the Kubernetes API design conventions 49
Understanding a CRD schema 51
Example Operator CRD 53

Working with other required resources 55
Designing a target reconciliation loop 57
Level- versus edge-based event triggering 58

Designing reconcile logic 58

Handling upgrades and downgrades 60
Using failure reporting 62
Reporting errors with logging 62
Reporting errors with status updates 63
Reporting errors with events 64

Summary 66

4

Developing an Operator with the Operator SDK

Technical requirements	70	Troubleshooting	97
Setting up your project	71	General Kubernetes resources	97
Defining an API	72	Operator SDK resources	99
Adding resource manifests	76	Kubebuilder resources	99
Additional manifests and BinData	81	Summary	100
Writing a control loop	89		

5

Developing an Operator – Advanced Functionality

Technical requirements	102	Implementing metrics reporting	114
Understanding the need for advanced functionality	102	Adding a custom service metric	115
Reporting status conditions	103	RED metrics	116
Operator CRD conditions	104	Implementing leader election	121
Using the OLM OperatorCondition	111	Adding health checks	125
		Summary	127

6

Building and Deploying Your Operator

Technical requirements	130	Redeploying the Operator with metrics	146
Building a container image	131	Key takeaways	147
Building the Operator locally	133	Troubleshooting	148
Building the Operator image with Docker	133	Makefile issues	148
		kind	148
Deploying in a test cluster	137	Docker	149
Pushing and testing changes	143	Metrics	151
Installing and configuring kube-prometheus	144	Additional errors	153
		Summary	153

Part 3: Deploying and Distributing Operators for Public Use

7

Installing and Running Operators with the Operator Lifecycle Manager

Technical requirements	**158**	
Understanding the OLM	**159**	
Installing the OLM in a Kubernetes cluster	159	
Interacting with the OLM	161	
Running your Operator	**163**	
Generating an Operator's bundle	163	
Exploring the bundle files	166	
Building a bundle image	169	
Pushing a bundle image	171	

Deploying an Operator bundle with the OLM	172
Working with OperatorHub	**173**
Installing Operators from OperatorHub	174
Submitting your own Operator to OperatorHub	178
Troubleshooting	**181**
OLM support	181
OperatorHub support	181
Summary	**183**

8

Preparing for Ongoing Maintenance of Your Operator

Technical requirements	**186**	
Releasing new versions of your Operator	**187**	
Adding an API version to your Operator	187	
Updating the Operator CSV version	206	
Releasing a new version on OperatorHub	208	
Planning for deprecation and backward compatibility	**209**	
Revisiting Operator design	209	
Starting small	210	
Iterating effectively	210	
Deprecating gracefully	211	

Complying with Kubernetes standards for changes	**211**
Removing APIs	212
API conversion	213
API lifetime	213
Aligning with the Kubernetes release timeline	**214**
Overview of a Kubernetes release	214
Start of release	216
Enhancements Freeze	216
Call for Exceptions	217
Code Freeze	218
Test Freeze	218

GA release/Code Thaw 219
Retrospective 220

Working with the Kubernetes community 220
Summary 221

9

Diving into FAQs and Future Trends

FAQs about the Operator Framework 224

What is an Operator? 224
What benefit do Operators provide to a Kubernetes cluster? 224
How are Operators different from other Kubernetes controllers? 224
What is the Operator Framework? 225
What is an Operand? 225
What are the main components of the Operator Framework? 225
What programming languages can Operators be written in? 226
What is the Operator Capability Model? 226

FAQs about Operator design, CRDs, and APIs 227

How does an Operator interact with Kubernetes? 227
What cluster resources does an Operator act on? 227
What is a CRD? 227
How is a CRD different from a CR object? 227
What Kubernetes namespaces do Operators run within? 228
How do users interact with an Operator? 228
How can you plan for changes early in an Operator's lifecycle? 228
How does an Operator's API relate to its CRD? 228

What are the conventions for an Operator API? 229
What is a structural CRD schema? 229
What is OpenAPI v3 validation? 229
What is Kubebuilder? 229
What is a reconciliation loop? 229
What is the main function of an Operator's reconciliation loop? 230
What are the two kinds of event triggering? 230
What is a ClusterServiceVersion (CSV)? 230
How can Operators handle upgrades and downgrades? 230
How can Operators report failures? 231
What are status conditions? 231
What are Kubernetes events? 231

FAQs about the Operator SDK and coding controller logic 231

What is the Operator SDK? 231
How can operator-sdk scaffold a boilerplate Operator project? 231
What does a boilerplate Operator project contain? 232
How can you create an API with operator-sdk? 232
What does a basic Operator API created with operator-sdk look like? 232
What other code is generated by operator-sdk? 232
What do Kubebuilder markers do? 232
How does the Operator SDK generate Operator resource manifests? 233

How else can you customize generated
Operator manifests?   233

What are go-bindata and go:embed?   233

What is the basic structure of a
control/reconciliation loop?   233

How does a control loop function
access Operator config settings?   234

What information does a status
condition report?   234

What are the two basic kinds of
metrics?   234

How can metrics be collected?   234

What are RED metrics?   234

What is leader election?   235

What are the two main strategies for
leader election?   235

What are health and ready checks?   235

**FAQs about OperatorHub and
the OLM**   **235**

What are the different ways to
compile an Operator?   235

How does a basic Operator SDK
project build a container image?   235

How can an Operator be deployed
in a Kubernetes cluster?   236

What is the OLM?   236

What benefit does running an
Operator with the OLM provide?   236

How do you install the OLM
in a cluster?   236

What does the operator-sdk olm
status command show?   236

What is an Operator bundle?   236

How do you generate a bundle?   237

What is a bundle image?   237

How do you build a bundle image?   237

How do you deploy a bundle with the
OLM?   237

What is OperatorHub?   237

How do you install an Operator from
OperatorHub?   238

How do you submit an Operator to
OperatorHub?   238

**Future trends in the Operator
Framework**   **239**

How do you release a new version of
an Operator?   239

When is it appropriate to add a new
API version?   239

How do you add a new API version?   239

What is an API conversion?   239

How do you convert between two
versions of an API?   240

What is a conversion webhook?   240

How do you add a conversion
webhook to an Operator?   240

What is kube-storage-version-migrator? 240

How do you update an Operator's CSV? 240

What are upgrade channels?   241

How do you publish a new version on
OperatorHub?   241

What is the Kubernetes deprecation
policy?   241

How can API elements be removed in
the Kubernetes deprecation policy?   241

How long are API versions generally
supported?   241

How long is the Kubernetes release
cycle?   242

What is Enhancements Freeze?   242

What is Code Freeze?   242

What is Retrospective?   242

How do Kubernetes community
standards apply to Operator
development?   242

Summary   **243**

10

Case Study for Optional Operators – the Prometheus Operator

A real-world use case	246	CRDs and APIs	254
Prometheus overview	247	Reconciliation logic	261
Installing and running Prometheus	247		
Configuring Prometheus	249	Operator distribution and development	266
Summarizing the problems with manual Prometheus	252	Updates and maintenance	271
		Summary	273
Operator design	254		

11

Case Study for Core Operator – Etcd Operator

Core Operators – extending the Kubernetes platform	276	Reconciliation logic	285
		Failure recovery	287
RBAC Manager	277		
The Kube Scheduler Operator	281	Stability and safety	288
The etcd Operator	282	Upgrading Kubernetes	290
		Summary	292
etcd Operator design	283		
CRDs	283		

Index

Other Books You May Enjoy

Preface

The emergence of Kubernetes as a standard platform for distributed computing has revolutionized the landscape of enterprise application development. Organizations and developers can now easily write and deploy applications with a cloud-native approach, scaling those deployments to meet the needs of them and their users. However, with that scale comes increasing complexity and a maintenance burden. In addition, the nature of distributed workloads exposes applications to increased potential points of failure, which can be costly and time-consuming to repair. While Kubernetes is a powerful platform on its own, it is not without its own challenges.

The Operator Framework has been developed specifically to address these pain points by defining a standard process for automating operational tasks in Kubernetes clusters. Kubernetes administrators and developers now have a set of APIs, libraries, management applications, and command-line tools to rapidly create controllers that automatically create and manage their applications (and even core cluster components). These controllers, called Operators, react to the naturally fluctuating state of a Kubernetes cluster to reconcile any deviation from the desired administrative stasis.

This book is an introduction to the Operator Framework for anyone who is interested in, but unfamiliar with, Operators and how they benefit Kubernetes users, with the goal of providing a practical lesson in designing, building, and using Operators. To that end, it is more than just a technical tutorial for writing Operator code (though it does walk through writing a sample Operator in Go). It is also a guide on intangible design considerations and maintenance workflows, offering a holistic approach to Operator use cases and development to guide you toward building and maintaining your own Operators.

Who this book is for

The target audience for this book is anyone who is considering using the Operator Framework for their own development, including engineers, project managers, architects, and hobbyist developers. The content in this book assumes some prior knowledge of basic Kubernetes concepts (such as Pods, ReplicaSets, and Deployments). However, there is no requirement for any prior experience with Operators or the Operator Framework.

What this book covers

Chapter 1, Introducing the Operator Framework, provides a brief introduction to the fundamental concepts and terminology that describe the Operator Framework.

Chapter 2, Understanding How Operators Interact with Kubernetes, provides sample descriptions of the ways that Operators function in a Kubernetes cluster, including not only the technical interactions but also descriptions of different user interactions.

Chapter 3, Designing an Operator – CRD, API, and Target Reconciliation, discusses high-level considerations to take into account when designing a new Operator.

Chapter 4, Developing an Operator with the Operator SDK, provides a technical walk - through of creating a sample Operator project in Go with the Operator SDK toolkit.

Chapter 5, Developing an Operator – Advanced Functionality, builds on the sample Operator project from the previous chapter to add more complex functionality.

Chapter 6, Building and Deploying Your Operator, demonstrates the processes for compiling and installing an Operator in a Kubernetes cluster by hand.

Chapter 7, Installing and Running Operators with the Operator Lifecycle Manager, provides an introduction to the Operator Lifecycle Manager, which helps to automate the deployment of Operators in a cluster.

Chapter 8, Preparing for Ongoing Maintenance of Your Operator, provides considerations for promoting the active maintenance of Operator projects, including how to release new versions and alignment with upstream Kubernetes release standards.

Chapter 9, Diving into FAQs and Future Trends, provides a distilled summary of the content from previous chapters, broken down into small FAQ-style sections.

Chapter 10, Case Study for Optional Operators – the Prometheus Operator, provides a demonstration of the Operator Framework concepts in a real-world example of an Operator used to manage applications.

Chapter 11, Case Study for Core Operator – Etcd Operator, provides an additional example of Operator Framework concepts applied to the management of core cluster components.

To get the most out of this book

It is assumed that you have at least a foundational understanding of basic Kubernetes concepts and terms due to the fact that the Operator Framework builds heavily on these concepts to serve its purpose. These include topics such as basic application deployment and a familiarity with command-line tools such as kubectl for interacting with Kubernetes clusters. While direct hands-on experience with these topics is not necessary, it will be helpful.

Software/hardware covered in the book	Operating system requirements
Kubernetes (kubectl)	Windows, macOS, or Linux
Git	Windows, macOS, or Linux
Docker	Windows, macOS, or Linux
Kind	Windows, macOS, or Linux

In addition, administrator access to a Kubernetes cluster is needed in order to complete all of the sample tasks in the book (for example, deploying an Operator in *Chapter 6, Building and Deploying Your Operator*). The chapters that require a Kubernetes cluster offer some options for creating disposable clusters and basic setup steps for doing so, but in order to focus on the main content, these sections intentionally do not go into thorough detail regarding cluster setup. It is strongly recommended to use a disposable cluster for all examples in order to avoid accidental damage to sensitive workloads.

If you are using the digital version of this book, we advise you to type the code yourself or access the code from the book's GitHub repository (a link is available in the next section). Doing so will help you avoid any potential errors related to the copying and pasting of code.

Download the example code files

You can download the example code files for this book from GitHub at https://github.com/PacktPublishing/The-Kubernetes-Operator-Framework-Book. If there's an update to the code, it will be updated in the GitHub repository.

We also have other code bundles from our rich catalog of books and videos available at https://github.com/PacktPublishing/. Check them out!

Code in Action

The Code in Action videos for this book can be viewed at https://bit.ly/3m5dlYa.

Download the color images

We also provide a PDF file that has color images of the screenshots and diagrams used in this book. You can download it here: `https://static.packt-cdn.com/downloads/9781803232850_ColorImages.pdf`.

Conventions used

There are a number of text conventions used throughout this book.

`Code in text`: Indicates code words in text, database table names, folder names, filenames, file extensions, pathnames, dummy URLs, user input, and Twitter handles. Here is an example: "This requires additional resources, such as `ClusterRoles` and `RoleBindings`, to ensure the Prometheus Pod has permission to scrape metrics from the cluster and its applications."

A block of code is set as follows:

```
apiVersion: monitoring.coreos.com/v1
kind: Prometheus
metadata:
  name: sample
spec:
  replicas: 2
```

When we wish to draw your attention to a particular part of a code block, the relevant lines or items are set in bold:

```
apiVersion: monitoring.coreos.com/v1
kind: ServiceMonitor
metadata:
  name: web-service-monitor
  labels:
    app: web
spec:
  selector:
    matchLabels:
      serviceLabel: webapp
```

Any command-line input or output is written as follows:

```
$ export BUNDLE_IMG=docker.io/sample/nginx-bundle:v0.0.2
$ make bundle-build bundle-push
$ operator-sdk run bundle docker.io/same/nginx-bundle:v0.0.2
```

Bold: Indicates a new term, an important word, or words that you see on screen. For instance, words in menus or dialog boxes appear in **bold**. Here is an example: "Clicking on the **Grafana Operator** tile opens up the information page for this specific Operator."

> **Tips or Important Notes**
> Appear like this.

Get in touch

Feedback from our readers is always welcome.

General feedback: If you have questions about any aspect of this book, email us at customercare@packtpub.com and mention the book title in the subject of your message.

Errata: Although we have taken every care to ensure the accuracy of our content, mistakes do happen. If you have found a mistake in this book, we would be grateful if you would report this to us. Please visit www.packtpub.com/support/errata and fill in the form.

Piracy: If you come across any illegal copies of our works in any form on the internet, we would be grateful if you would provide us with the location address or website name. Please contact us at copyright@packt.com with a link to the material.

If you are interested in becoming an author: If there is a topic that you have expertise in and you are interested in either writing or contributing to a book, please visit authors.packtpub.com.

Share Your Thoughts

Once you've read *The Kubernetes Operator Framework Book*, we'd love to hear your thoughts! Scan the QR code below to go straight to the Amazon review page for this book and share your feedback.

https://packt.link/r/1803232854

Your review is important to us and the tech community and will help us make sure we're delivering excellent quality content.

Part 1: Essentials of Operators and the Operator Framework

In this section, you will achieve a basic understanding of the history and purpose of Kubernetes Operators. The fundamental concepts of the Operator Framework will be introduced and you will learn how Operators function in a Kubernetes cluster. This will set the groundwork for more complex concepts, which will be introduced later.

This section comprises the following chapters:

- *Chapter 1, Introducing the Operator Framework*
- *Chapter 2, Understanding How Operators Interact with Kubernetes*

1
Introducing the Operator Framework

Managing a Kubernetes cluster is hard. This is partly due to the fact that any microservice architecture is going to be inherently based on the interactions of many small components, each introducing its own potential point of failure. There are, of course, many benefits to this type of system design, such as graceful error handling thanks to the separation of responsibilities. However, diagnosing and reconciling such errors requires significant engineering resources and a keen familiarity with an application's design. This is a major pain point for project teams who migrate to the Kubernetes platform.

The Operator Framework was introduced to the Kubernetes ecosystem to address these problems. This chapter will go over a few general topics to give a broad overview of the Operator Framework. The intent is to provide a brief introduction to the Operator Framework, the problems it solves, how it solves them, and the tools and patterns it provides to users. This will highlight key takeaways for the goals and benefits of using Operators to help administrata a Kubernetes cluster. These topics include the following:

- Managing clusters without Operators
- Introducing the Operator Framework

- Developing with the Operator **software development kit (SDK)**
- Managing Operators with the **Operator Lifecycle Manager (OLM)**
- Distributing Operators on `OperatorHub.io`
- Defining Operator functions with the Capability Model
- Using Operators to manage applications

Technical requirements

This chapter does not have any technical requirements because we will only be covering general topics. In later chapters, we will discuss these various topics in depth and include technical prerequisites for following along with them.

The Code in Action video for this chapter can be viewed at: `https://bit.ly/3GKJfmE`

Managing clusters without Operators

Kubernetes is a powerful microservice container orchestration platform. It provides many different controllers, resources, and design patterns to cover almost any use case, and it is constantly growing. Because of this, applications designed to be deployed on Kubernetes can be very complex.

When designing an application to use microservices, there are a number of concepts to be familiar with. In Kubernetes, these are mainly the native **application programming interface (API)** resource objects included in the core platform. Throughout this book, we will assume a foundational familiarity with the common Kubernetes resources and their functions.

These objects include Pods, Replicas, Deployments, Services, Volumes, and more. The orchestration of any microservice-based cloud application on Kubernetes relies on integrating these different concepts to weave a coherent whole. This orchestration is what creates a complexity that many application developers struggle to manage.

Demonstrating on a sample application

Take, for example, a simple web application that accepts, processes, and stores user input (such as a message board or chat server). A good, containerized design for an application such as this would be to have one Pod presenting the frontend to the user and a second backend Pod that accepts the user's input and sends it to a database for storage.

Of course, you will then need a Pod running the database software and a Persistent Volume to be mounted by the database Pod. These three Pods will benefit from Services to communicate with each other, and they will also need to share some common environment variables (such as access credentials for the database and environment variables to tweak different application settings).

Here is a diagram of what a sample application of this sort could look like. There are three Pods (frontend, backend, and database), as well as a Persistent Volume:

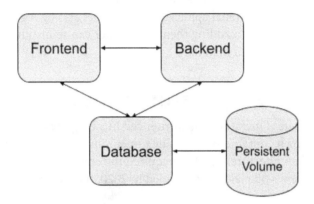

Figure 1.1 – Simple application diagram with three Pods and a Persistent Volume

This is just a small example, but it's already evident how even a simple application can quickly involve tedious coordination between several moving parts. In theory, these discrete components will all continue to function cohesively as long as each individual component does not fail. But what about when a failure does occur somewhere in the application's distributed design? It is never wise to assume that an application's valid state will consistently remain that way.

Reacting to changing cluster states

There are a number of reasons a cluster state can change. Some may not even technically be considered a failure, but they are still changes of which the running application must be aware. For example, if your database access credentials change, then that update needs to be propagated to all the Pods that interact with it. Or, a new feature is available in your application that requires tactful rollout and updated settings for the running workloads. This requires manual effort (and, more importantly, time), along with a keen understanding of the application architecture.

Time and effort are even more critical in the case of an unexpected failure. These are the kinds of problems that the Operator Framework addresses automatically. If one of the Pods that make up this application hits an exception or the application's performance begins to degrade, these scenarios require intervention. That means a human engineer must not only know the details of the deployment, but they must also be on-call to maintain uptime at any hour.

There are additional components that can help administrators monitor the health and performance of their applications, such as metrics aggregation servers. However, these components are essentially additional applications that must also be regularly monitored to make sure they are working, so adding them to a cluster can reintroduce the same issues of managing an application manually.

Introducing the Operator Framework

The concept of Kubernetes Operators was introduced in a blog post in 2016 by CoreOS. CoreOS created their own container-native Linux operating system that was optimized for the needs of cloud architecture. Red Hat acquired the company in 2018, and while the CoreOS operating system's official support ended in 2020, their Operator Framework has thrived.

The principal idea behind an Operator is to automate cluster and application management tasks that would normally be done manually by a human. This role can be thought of as an automated extension of support engineers or **development-operations** (**DevOps**) teams.

Most Kubernetes users will already be familiar with some of the design patterns of Operators, even if they have never used the Operator Framework before. This is because Operators are a seemingly complicated topic, but ultimately, they are not functionally much different than many of the core components that already automate most of a Kubernetes cluster by default. These components are called controllers, and at its core, any Operator is essentially just a controller.

Exploring Kubernetes controllers

Kubernetes itself is made up of many default controllers. These controllers maintain the desired state of the cluster, as set by users and administrators. Deployments, ReplicaSets, and Endpoints are just a few examples of cluster resources that are managed by their own controllers. Each of these resources involves an administrator declaring the desired cluster state, and it is then the controller's job to maintain that state. If there is any deviation, the controller must act to resolve what they control.

These controllers work by monitoring the current state of the cluster and comparing it to the desired state. One example is a ReplicaSet with a specification to maintain three replicas of a Pod. Should one of the replicas fail, the ReplicaSet quickly identifies that there are now only two running replicas. It then creates a new Pod to bring stasis back to the cluster.

In addition, these core controllers are collectively managed by the **Kube Controller Manager**, which is another type of controller. It monitors the state of controllers and attempts to recover from errors if one fails or reports the error for human intervention if it cannot automatically recover. So, it is even possible to have controllers that manage other controllers.

In the same way, Kubernetes Operators put the development of operational controllers in the hands of users. This provides administrators with the flexibility to write a controller that can manage any aspect of a Kubernetes cluster or custom application. With the ability to define more specific logic, developers can extend the main benefits of Kubernetes to the unique needs of their own applications.

The Operators that are written following the guidelines of the Operator Framework are designed to function very similarly to native controllers. They do this by also monitoring the current state of the cluster and acting to reconcile it with the desired state. Specifically, an Operator is tailored to a unique workload or component. The Operator then knows how to interact with that component in various ways.

Knowing key terms for Operators

The component that is managed by an Operator is its **Operand**. An Operand is any kind of application or workload whose state is reconciled by an Operator. Operators can have many Operands, though most Operators manage—at most—just a few (usually just one). The key distinction is that Operators exist to manage Operands, where the Operator is a meta-application in the architectural design of the system.

Operands can be almost any type of workload. While some Operators manage application deployments, many others deploy additional, optional cluster components offering meta-functionality such as database backup and restoration. Some Operators even make core native Kubernetes components their Operands, such as etcd. So, an Operator doesn't even need to be managing your own workloads; they can help with any part of a cluster.

No matter what the Operator is managing, it must provide a way for cluster administrators to interact with it and configure settings for their application. An Operator exposes its configuration options through a **Custom Resource**.

Custom Resources are created as API objects following the constraints of a matching **CustomResourceDefinition (CRD)**. CRDs are themselves a type of native Kubernetes object that allows users and administrators to extend the Kubernetes platform with their own resource objects beyond what is defined in the core API. In other words, while a Pod is a built-in native API object in Kubernetes, CRDs allow cluster administrators to define *MyOperator* as another API object and interact with it the same way as native objects.

Putting it all together

The Operator Framework strives to define an entire ecosystem for Operator development and distribution. This ecosystem comprises three pillars that cover the coding, deployment, and publishing of Operators. They are the Operator SDK, OLM, and OperatorHub.

These three pillars are what have made the Operator Framework so successful. They transform the framework from just development patterns to an encompassing, iterative process that spans the entire lifecycle of an Operator. This helps support the contract between Operator developers and users to provide consistent industry standards for their software.

The lifecycle of an Operator begins with development. To help with this, the Operator SDK exists to guide developers in the first steps of creating an Operator. Technically, an Operator does not have to be written with the Operator SDK, but the Operator SDK provides development patterns to significantly reduce the effort needed to bootstrap and maintain an Operator's source code.

While coding and development are certainly important parts of creating an Operator, any project's timeline does not end once the code is compiled. The Operator Framework community recognized that a coherent ecosystem of projects must offer guidance beyond just the initial development stage. Projects need consistent methods for installation, and as software evolves, there is a need to publish and distribute new versions. OLM and OperatorHub help users to install and manage Operators in their cluster, as well as share their Operators in the community.

Finally, the Operator Framework provides a scale of Operator functionality called the Capability Model. The Capability Model provides developers with a way to classify the functional abilities of their Operator by answering quantifiable questions. An Operator's classification, along with the Capability Model, gives users information about what they can expect from the Operator.

Together, these three pillars establish the basis of the Operator Framework and form the design patterns and community standards that distinguish Operators as a concept. Along with the Capability Model, this standardized framework has led to an explosion in the adoption of Operators in Kubernetes.

At this point, we have discussed a brief introduction to the core concepts of the Operator Framework. In contrast with a Kubernetes application managed without an Operator, the pillars of the Operator Framework address problems met by application developers. This understanding of the core pillars of the Operator Framework will set us up for exploring each of them in more depth.

Developing with the Operator SDK

The first pillar of the Operator Framework is the Operator SDK. As with any other software development toolkit, the Operator SDK provides packaged functionality and design patterns as code. These include predefined APIs, abstracted common functions, code generators, and project scaffolding tools to easily start an Operator project from scratch.

The Operator SDK is primarily written in Go, but its tooling allows Operators to be written using Go code, Ansible, or Helm. This gives developers the ability to write their Operators from the ground up by coding the CRDs and reconciliation logic themselves, or by taking advantage of automated deployment tools provided by Ansible and Helm to generate their APIs and reconciliation logic depending on their needs.

Developers interact with the Operator SDK through its `operator-sdk` command-line binary. The binary is available on Homebrew for Mac and is also available directly from the Operator Framework GitHub repository (`https://github.com/operator-framework/operator-sdk`) as a release, where it can also be compiled from source.

Whether you are planning to develop an Operator with **Go, Ansible,** or **Helm**, the Operator SDK binary provides commands to initialize the boilerplate project source tree. These commands include `operator-sdk init` and `operator-sdk create api`. The first command initializes a project's source directory with boilerplate Go code, dependencies, hack scripts, and even a `Dockerfile` and `Makefile` for compiling the project.

Creating an API for your Operator is necessary to define the CRD required to interact with the Operator once it is deployed in a Kubernetes cluster. This is because CRDs are backed by API type definitions written in Go code. The CRD is generated from these code definitions, and the Operator has logic built in to translate between CRD and Go representations of the object. Essentially, CRDs are how users interact with Operators, and Go code is how the Operator understands the settings. CRDs also add benefits such as structural validation schemas to automatically validate inputs.

The Operator SDK binary has flags to specify the name and version of the API. It then generates the API types as Go code and corresponding **YAML Ain't Markup Language** (**YAML**) files based on best-practice standard definitions. However, you are free to modify the definitions of your API in whichever way you choose.

If we were to initialize a basic Operator for an application such as the one first demonstrated at the start of this chapter, the steps would be relatively simple. They would look like this:

```
$ mkdir sample-app
$ cd sample-app/
$ operator-sdk init --domain mysite.com --repo github.com/
sample/simple-app
$ operator-sdk create api --group myapp --version v1alpha1
--kind WebApp --resource -controller
$ ls
total 112K
drwxr-xr-x    15 mdame  staff   480 Nov 15 17:00 .
drwxr-xr-x+  270 mdame  staff  8.5K Nov 15 16:48 ..
drwx------     3 mdame  staff    96 Nov 15 17:00 api
drwxr-xr-x     3 mdame  staff    96 Nov 15 17:00 bin
drwx------    10 mdame  staff   320 Nov 15 17:00 config
drwx------     4 mdame  staff   128 Nov 15 17:00 controllers
drwx------     3 mdame  staff    96 Nov 15 16:50 hack
-rw-------     1 mdame  staff   129 Nov 15 16:50 .dockerignore
-rw-------     1 mdame  staff   367 Nov 15 16:50 .gitignore
-rw-------     1 mdame  staff   776 Nov 15 16:50 Dockerfile
-rw-------     1 mdame  staff  8.7K Nov 15 16:51 Makefile
-rw-------     1 mdame  staff   422 Nov 15 17:00 PROJECT
-rw-------     1 mdame  staff   218 Nov 15 17:00 go.mod
-rw-r--r--     1 mdame  staff   76K Nov 15 16:51 go.sum
-rw-------     1 mdame  staff  3.1K Nov 15 17:00 main.go
```

After this, you would go on to develop the logic of the Operator based on the method you choose. If that's to write Go code directly, it would start by modifying the *.go files in the project tree. For Ansible and Helm deployments, you would begin working on the Ansible roles or Helm chart for your project.

Finally, the Operator SDK binary provides a set of commands to interact with OLM. These include the ability to install OLM in a running cluster, but also install and manage specific Operators within OLM.

Managing Operators with OLM

OLM is the second pillar of the Operator Framework. Its purpose is to facilitate the deployment and management of Operators in a Kubernetes cluster. It is a component that runs within a Kubernetes cluster and provides several commands and features for interacting with Operators.

OLM is primarily used for the installation and upgrade of Operators—this includes fetching and installing any dependencies for those Operators. Users interact with OLM via commands provided by the Operator SDK binary, the Kubernetes command-line tool (kubectl), and declarative YAML.

To get started, OLM can be initialized in a cluster with the following command:

```
$ operator-sdk olm install
```

Besides installing Operators, OLM can also make Operators that are currently installed discoverable to users on the cluster. This provides a catalog of already installed Operators available to cluster users. Also, by managing all the known Operators in the cluster, OLM can watch for conflicting Operator APIs and settings that would destabilize the cluster.

Once an Operator's Go code is compiled into an image, it is ready to be installed into a cluster with OLM running. Technically, OLM is not required to run an Operator in any cluster. For example, it is completely possible to deploy an Operator manually in the cluster, just as with any other container-based application. However, due to the advantages and security measures described previously (including its ability to install Operators and its awareness of other installed Operators), it is highly recommended to use OLM to manage cluster Operators.

When developing an Operator, the image is compiled into a **bundle**, and that bundle is installed via OLM. The bundle consists of several YAML files that describe the Operator, its CRD, and its dependencies. OLM knows how to process this bundle in its standardized format to properly manage the Operator in a cluster.

Compiling an Operator's code and deploying it can be done with commands such as the ones shown next. The first command shown in the following code snippet builds the bundle of YAML manifests that describe the Operator. Then, it passes that information to OLM to run the Operator in your cluster:

```
$ make bundle ...
$ operator-sdk run bundle ...
```

Later chapters will demonstrate exactly how to use these commands and what they do, but the general idea is that these commands first compile the Operator's Go code into an image and a deployable format that's understandable by OLM. But OLM isn't the only part of the Operator Framework that consumes an Operator's bundle—much of the same information is used by OperatorHub to provide information on an Operator.

Once an Operator has been compiled into its image, OperatorHub exists as a platform to share and distribute those images to other users.

Distributing Operators on OperatorHub.io

The final core component of the Operator Framework is OperatorHub.io. As a major open source project, the Operator Framework ecosystem is built on the open sharing and distribution of projects. Therefore, OperatorHub powers the growth of Operators as a Kubernetes concept.

OperatorHub is an open catalog of Operators published and managed by the Kubernetes community. It serves as a central index of freely available Operators, each contributed by developers and organizations. You can see an overview of the OperatorHub.io home page in the following screenshot:

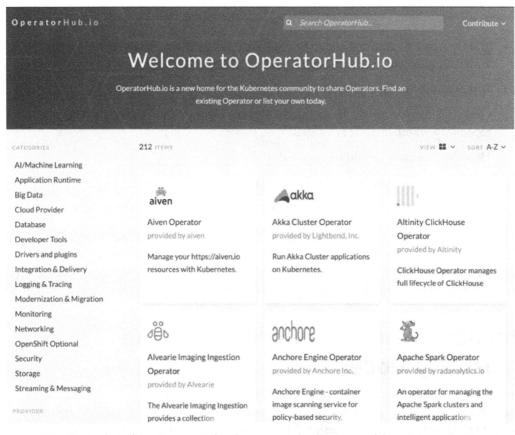

Figure 1.2 – Screenshot of the OperatorHub.io home page, showing some of the most popular Operators

The process for submitting an Operator to OperatorHub for indexing has been standardized to ensure the consistency and compatibility of Operators with OLM. New Operators are reviewed by automated tooling for compliance with this standard definition of an Operator. The process is mainly handled through the open source GitHub repository that provides the backend of OperatorHub. However, OperatorHub does not provide any assistance with the ongoing maintenance of an Operator, which is why it is important for Operator developers to share links to their own open source repositories and contact information where users can report bugs and contribute themselves.

Preparing an Operator for submission to OperatorHub involves generating its bundle and associated manifests. The submission process primarily relies on the Operator's **Cluster Service Version (CSV)**. The CSV is a YAML file that provides most of the metadata to OLM and OperatorHub about your Operator. It includes general information such as the Operator's name, version, and keywords. However, it also defines installation requirements (such as **role-based access control** (**RBAC**) permissions), CRDs, APIs, and additional cluster resource objects owned by the Operator.

The specific sections of an Operator's CSV include the following:

- The Operator's name and version number, as well as a description of the Operator and its display icon in Base64-encoded image format

- Annotations for the Operator

- Contact information for the maintainers of the Operator and the open source repository where its code is located

- How the Operator should be installed in the cluster

- Example configurations for the Operator's CRD

- Required CRDs and other resources and dependencies that the Operator needs to run

Because of all the information that it covers, the Operator CSV is usually very long and takes time to prepare properly. However, a well-defined CSV helps an Operator reach a much wider audience. Details of Operator CSVs will be covered in a later chapter.

Defining Operator functions with the Capability Model

The Operator Framework defines a Capability Model (`https://operatorframework.io/operator-capabilities/`) that categorizes Operators based on their functionality and design. This model helps to break down Operators based on their maturity, and also describes the extent of an Operator's interoperability with OLM and the capabilities users can expect when using the Operator.

The Capability Model is divided into five hierarchical levels. Operators can be published at any one of these levels and, as they grow, may evolve and graduate from one level to the next as features and functionality are added. In addition, the levels are cumulative, with each level generally encompassing all features of the levels below it.

The current level of an Operator is part of the CSV, and this level is displayed on its OperatorHub listing. The level is based on somewhat subjective yet guided criteria and is purely an informational metric.

Each level has specific functionalities that define it. These functionalities are broken down into *Basic Install, Seamless Upgrades, Full Lifecycle, Deep Insights*, and *Auto Pilot*. The specific levels of the Capability Model are outlined here:

1. **Level I—Basic Install**: This level represents the most basic of Operator capabilities. At *Level I*, an Operator is only capable of installing its Operand in the cluster and conveying the status of the workload to cluster administrators. This means that it can set up the basic resources required for an application and report when those resources are ready to be used by the cluster.

   At *Level I*, an Operator also allows for simple configuration of the Operand. This configuration is specified through the Operator's Custom Resource. The Operator is responsible for reconciling the configuration specifications with the running Operand workload. However, it may not be able to react if the Operand reaches a failed state, whether due to malformed configuration or outside influence.

   Going back to our example web application from the start of the chapter, a *Level I* Operator for this application would handle the basic setup of the workloads and nothing else. This is good for a simple application that needs to be quickly set up on many different clusters, or one that should be easily shared with users for them to install themselves.

2. **Level II—Seamless Upgrades**: Operators at *Level II* offer the features of basic installation, with added functionality around upgrades. This includes upgrades for the Operand but also upgrades for the Operator itself.

   Upgrades are a critical part of any application. As bug fixes are implemented and more features are added, being able to smoothly transition between versions helps ensure application uptime. An Operator that handles its own upgrades can either upgrade its Operand when it upgrades itself or manually upgrade its Operand by modifying the Operator's Custom Resource.

   For seamless upgrades, an Operator must also be able to upgrade older versions of its Operand (which may exist because they were managed by an older version of the Operator). This kind of backward compatibility is essential for both upgrading to newer versions and handling rollbacks (for example, if a new version introduces a high-visibility bug that can't wait for an eventual fix to be published in a patch version).

Our example web application Operator could offer the same set of features. This means that if a new version of the application were released, the Operator could handle upgrading the deployed instances of the application to the newer version. Or, if changes were made to the Operator itself, then it could manage its own upgrades (and later upgrade the application, regardless of version skew between Operator and Operand).

3. **Level III—Full Lifecycle**: *Level III* Operators offer at least one out of a list of Operand lifecycle management features. Being able to offer management during the Operand's lifecycle implies that the Operator is more than just passively operating on a workload in a *set and forget* fashion. At Level III, Operators are actively contributing to the ongoing function of the Operand.

The features relevant to the lifecycle management of an Operand include the following:

- The ability to create and/or restore backups of the Operand.

- Support for more complex configuration options and multistep workflows.

- Failover and failback mechanisms for **disaster recovery** (**DR**). When the Operator encounters an error (either in itself or the Operand), it needs to be able to either re-route to a backup process (fail over) or roll the system back to its last known functioning state (fail back).

- The ability to manage clustered Operands, and—specifically—support for adding and removing members to and from Operands. The Operator should be capable of considering quorum for Operands that run multiple replicas.

- Similarly, support for scaling an Operand with worker instances that operate with read-only functionality.

Any Operator that implements one or more of these features can be considered to be at least a Level III Operator. The simple web application Operator could take advantage of a few of these, such as DR and scaling. As the user base grows and resources demands increase, an administrator could instruct the Operator to scale the application with additional replica Pods to handle the increased load.

Should any of the Pods fail during this process, the Operator would be smart enough to know to fail over to a different Pod or cluster zone entirely. Alternatively, if a new version of the web app was released that introduced an unexpected bug, the Operator could be aware of the previous successful version and provide ways to downgrade its Operand workloads if an administrator noticed the error.

4. **Level IV—Deep Insights**: While the previous levels focus primarily on Operator features as they relate to functional interaction with the application workload, Level IV emphasizes monitoring and metrics. This means an Operator is capable of providing measurable insights to the status of both itself and its Operand.

   Insights may be seen as less important from a development perspective relative to features and bug fixes, but they are just as critical to an application's success. Quantifiable reports about an application's performance can drive ongoing development and highlight areas that need improvement. Having a measurable system to push these efforts allows a way to scientifically prove or disprove which changes have an effect.

   Operators most commonly provide their insights in the form of metrics. These metrics are usually compatible with metrics aggregation servers such as Prometheus. (Interestingly enough, Red Hat publishes an Operator for Prometheus that is a Level IV Operator. That Operator is available on OperatorHub at `https://operatorhub.io/operator/prometheus`.)

   However, Operators can provide insights through other means as well. These include alerts and Kubernetes Events. Events are built-in cluster resource objects that are used by core Kubernetes objects and controllers.

   Another key insight that Level IV Operators report is the performance of the Operator and Operand. Together, these insights help inform administrators about the health of their clusters.

   Our simple web application Operator could provide insights about the performance of the Operand. Requests to the app would provide information about the current and historic load on the cluster. Additionally, since the Operator can identify failed states at this point, it could trigger an alert when the application is unhealthy. Many alerts would indicate
   a reliability issue that would gain the attention of an administrator.

5. **Level V—Auto Pilot**: Level V is the most sophisticated level for Operators. It includes Operators that offer the highest capabilities, in addition to the features in all four previous levels. This level is called *Auto Pilot* because the features that define it focus on being able to run almost entirely autonomously. These capabilities include Auto Scaling, Auto-Healing, Auto-Tuning, and Abnormality Detection.

    Auto Scaling is the ability for an Operator to detect the need to scale an application up or down based on demand. By measuring the current load and performance, an Operator can determine whether more or fewer resources are necessary to satisfy the current usage. Advanced Operators can even try to predict the need to scale based on current and past data.

    Auto-Healing Operators can react to applications that are reporting unhealthy conditions and work to correct them (or, at least, prevent them from getting any worse). When an Operand is reporting an error, the Operator should take reactive steps to rectify the failure. In addition, Operators can use current metrics to proactively prevent an Operand from transitioning to a failure state.

    Auto-Tuning means that an Operator can dynamically modify an Operand for peak performance. This involves tuning the settings of an Operand automatically. It can even include complex operations such as shifting workloads to entirely different nodes that are better suited than their current nodes.

    Finally, Abnormality Detection is the capability of an Operator to identify suboptimal or off-pattern behavior in an Operand. By measuring performance, an Operator has a picture of the application's current and historical levels of functioning. That data can be compared to a manually defined minimum expectation or used to dynamically inform the Operator of that expectation.

    All of these features are heavily dependent upon the use of metrics to automatically inform the Operator of the need to act upon itself or its Operand. Therefore, a Level V Operator is an inherent progression from Level IV, which is the level at which an Operator exposes advanced metrics.

    At Level V, the simple web application Operator would manage most of the aspects of the application for us. It has insights into the current number of requests, so it can scale up copies of the app on demand. If this scaling starts to cause errors (for example, too many concurrent database calls), it can identify the number of failing Pods and prevent further scaling. It would also attempt to modify parameters of the web app (such as request timeouts) to help rectify the situation and allow the auto-scaling to proceed. When the load peak subsided, the Operator would then automatically scale down the application to its baseline service levels.

Levels I and II (*Basic Install* and *Seamless Upgrades*) can be used with the three facets of the Operator SDK: Helm, Ansible, and Go. However, Level III and above (*Full Lifecycle*, *Deep Insights*, and *Auto Pilot*) are only possible with Ansible and Go. This is because the functionality at these higher levels requires more intricate logic than what is available through Helm charts alone.

We have now explained the three main pillars of the Operator Framework: Operator SDK, OLM, and OperatorHub. We learned how each contributes different helpful features to the development and usage of Operators. We also learned about the Capability Model, which serves as a reference for the different levels of functionality that Operators can have. In the next section, we'll apply this knowledge to a sample application.

Using Operators to manage applications

Clearly, working with Operators involves more than simply reconciling a cluster state. The Operator Framework is an encompassing platform for Kubernetes developers and users to solve unique problems, which makes Kubernetes so flexible.

Cluster administrators' first step in the Operator Framework is usually either with the Operator SDK, to develop their own Operator if there are no existing Operators that address their needs, or OperatorHub if there are.

Summarizing the Operator Framework

When developing an Operator from scratch, there are three choices for development methods: Go, Ansible, or Helm. However, using Ansible or Helm alone will ultimately limit the Operator's capabilities to the most basic levels of functionality.

If the developer wishes to share their Operator, they will need to package it into the standard manifest bundle for OperatorHub. Following a review, their Operator will be available publicly for other users to download and install in their own clusters.

OLM then makes it easy for users to launch Operators in a cluster. These Operators can be sourced from OperatorHub or written from scratch. Either way, OLM makes Operator installation, upgrades, and management much easier. It also provides several stability benefits when working with many Operators. You can see the relationship between the three services in the following diagram:

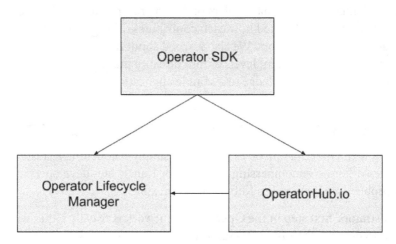

Figure 1.3 – The relationship between the Operator SDK, OperatorHub, and OLM

Each of these pillars provides distinct functions that aid in the development of Operators. Together, they comprise the foundation of the Operator Framework. Utilization of these pillars is the key distinguishing factor between an Operator and a normal Kubernetes controller. To summarize, while every Operator is essentially a controller, not every controller is an Operator.

Applying Operator capabilities

Revisiting the first example in this chapter, the idea of a simple application with three Pods and a Persistent Volume was examined without Operator management. This application relied on optimistic uptime and future-proof design to run continuously. In real-world deployments, these ideas are unfortunately unreasonable. Designs evolve and change, and unforeseeable failures bring applications down. But how could an Operator help this app persist in an unpredictable world?

By defining a single declarative configuration, this Operator could control various settings of the application deployment in one spot. This is the reason Operators are built on CRDs. These custom objects allow developers and users to easily interact with their Operators just as if they were native Kubernetes objects. So, the first step in writing an Operator to manage our simple web application would be to define a basic code structure with a CRD that has all the settings we think we'll need. Once we have done this, the new diagram of our application will look like this:

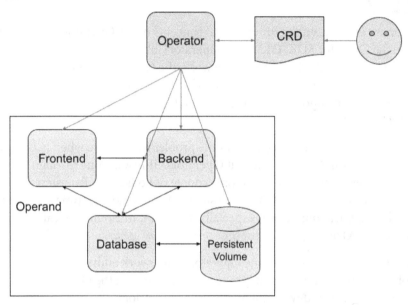

Figure 1.4 – In the new app layout, the cluster administrator only interacts with the Operator; the Operator then manages the workload

This shows how the details of the Operand deployment have been abstracted away from requiring manual administrator control, and the great part about CRDs is that more settings can be added in later versions of the Operator as our app grows. A few examples of settings to start with could be these:

- Database access information
- Application behavior settings
- Log level

While writing our Operator code, we'll also want to write logic for things such as metrics, error handling, and reporting. The Operator can also start to bidirectionally communicate with the Operand. This means that not only can it install and update the Operand, but it can receive communication back from the Operand about its status and report that as well.

Summary

In this chapter, we introduced the fundamental concepts of the Operator Framework. These include the Operator SDK, OLM, and OperatorHub. In addition to the development and distribution pillars of the Operator Framework, the Capability Model provides an additional tool for measuring the functionality of an Operator. Throughout this book, we will be exploring these components in deeper detail to get a hands-on understanding of how they actually work.

We began this chapter by examining some of the problems that arise when manually managing applications and clusters without Operators. This was done through the lens of a simple generic web application based on a couple of Pods and a Persistent Volume. The main difficulties in managing something such as this include the time and resources required to debug applications. This is especially important in cloud applications, where **high availability** (**HA**) and consistent uptime are top priorities.

We then looked at how each pillar of the Operator Framework addresses the biggest difficulties of application management. These pillars begin with the Operator SDK, which streamlines Operator development. This allows developers to begin iterating on automated reconciliation logic to get their Operators written quickly. It also provides commands to interact with OLM, which is the next pillar of the framework.

OLM exists to help administrators install and curate Operators within a cluster. It provides dependency management and notifies administrators of conflicting APIs to promote cluster stability. It also serves as a local catalog of installed Operators, which is useful for users on the cluster.

Next, we examined OperatorHub and its role in the broader open source Kubernetes community. As an open index of freely available Operators, OperatorHub serves to promote the adoption and maintenance of Operators. It consumes the same manifests as OLM to provide a standardized set of metadata about each Operator to users.

Finally, the Capability Model summarizes the maturity of an Operator based on the functionality it provides. This is helpful to users, but it also serves as a convenient roadmap for developers to plan features for their Operators.

To summarize each of these components, we revisited the original application example presented in the first section. We showed that, with an Operator in place to manage the application, cluster administrators do not need to be keenly aware of the architectural details of the app to keep it running. Instead, this information and controls are abstracted away behind the Operator's interface.

With all of this in mind, we move on to the next chapters to explore each of these topics in depth. We will also be following detailed examples to build our own sample Operator. In the next chapter, we'll begin looking at the important concepts of designing an Operator based on its interactions with a Kubernetes cluster.

2
Understanding How Operators Interact with Kubernetes

Now that we understand *what* kinds of things an Operator does (and *why* it does them), we can begin to explore *how* it performs these tasks. After identifying a use case for an Operator, taking the steps to lay out its technical design is the next step of the process. Though this is the period before any actual coding begins, it is still an essential part of development. This is the standard approach for almost any software project, and in this chapter, we will frame it in the context of Kubernetes.

During this planning phase, there are several factors to consider and questions to answer that will help guide the Operator's design. These factors are both technical and organic since your Operator needs to interact with not only the Kubernetes resources in your cluster but also the human resources of your engineers and administrators.

This chapter will explain some of the key considerations that should factor into your Operator's design. First, we will provide an introduction to some of the native Kubernetes components and resources that many Operators interact with. By looking at the ways that Operators consume these resources and the use cases for doing so, we can start looking at some of the patterns for functional Operator design. Next, we'll look at designing an Operator while keeping its unique user base in mind and the ways that the design can benefit users. Finally, we'll discuss some of the best practices for futureproofing an Operator for ongoing developments.

In this chapter, we will cover the following topics:

- Interacting with Kubernetes cluster resources
- Identifying your users and maintainers
- Designing beneficial features for your Operator
- Planning for evolving changes in your Operator

The goal of this chapter is to help guide you through the early design process of an Operator. Poorly planning an Operator's design can lead to many changes and updates being needed during the Operator's life cycle. This puts stress on engineering resources, but it can also confuse and frustrate users. That is why most of this chapter will focus on non-technical specifics. However, we do need to know some baseline technical interactions to shape these for our users. By the end of this chapter, you may find that you are not currently able to address all of the pre-design concerns that are described here. This is fine – the important part is to understand certain concepts that you must be aware of as you head into development.

Interacting with Kubernetes cluster resources

Before you decide on how to design an Operator with users' experiences in mind, it is important to understand what an Operator is capable of from a technical standpoint. Having a mindset for the specific capabilities of the Operator's code base will help guide the rest of the design process around what is truly possible. Otherwise, trying to define the scope of an Operator based solely on user requests without considering feasibility could lead to over-promising and under-delivering on functionality and usability.

The possibilities of any Operator are inherently limited by the underlying features of the Kubernetes platform. This platform is composed of different native cluster resources, some of which you may already be familiar with. This section will examine the most common resources that Operators work with and explain how they can and should use them. When you're developing an Operator, these resources will usually be consumed via the Kubernetes client libraries, which allow any application to interact with cluster resources.

Pods, ReplicaSets, and Deployments

Perhaps the most basic unit of a Kubernetes cluster architecture is the **Pod**. These objects represent one or more running containers. Under the hood, a Pod is essentially a definition of container images that the **Kubelet** can use to instruct the container runtime where and how to run certain containers. Applications are deployed on Kubernetes as Pods, Kubernetes itself is made up of many system Pods, and Operators are deployed as Pods as well.

While Pods are essential to a Kubernetes cluster, they are usually too atomic to manage manually. It is also often necessary to run multiple copies of an application, which is where **ReplicaSets** come in. The role of ReplicaSets is to define several replicas for a certain template of Pods.

However, ReplicaSets are also limited in their abilities. They simply maintain the number of replicas of a Pod on the cluster. **Deployments** add even more functionality to this by encompassing ReplicaSets and defining further controls such as rollout strategies and revision management for rollbacks.

From an Operator developer's perspective, Deployments are usually the most important resources to interact with. The advanced mechanics of Deployments provide flexibility for an application's runtime. These mechanics can be abstracted or restricted from users since they will only be interacting with the Operator through its CRD (rather than using the Deployment directly). However, the flexibility is still there to add it later or automatically program it into the Operator's reconciliation logic.

The following diagram shows how the encapsulation of application workload Pods within ReplicaSets and Deployments can still be managed by an Operator. In this example, the Operator only cares about the Deployment, but the status of that Deployment reflects the health of the actual workloads within it:

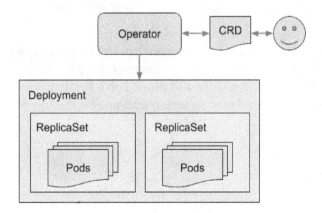

Figure 2.1 – An Operator managing a Deployment with two ReplicaSets, each with several Pods

An Operator will usually be installed and managed by a Deployment. The Deployment then provides a good owner reference of all the resource components that comprise the Operator for garbage collection. This is useful for upgrading and uninstalling the Operator.

Although it is not uncommon for an Operator to directly manage some Pods or ReplicaSets, there are certainly tradeoffs. For one, such an Operator's design would lead to a simpler application architecture at the expense of the capabilities and conveniences of using Deployments. So, when you're deciding if your Operator will directly manage Pods or a Deployment, it is important to consider the intent of the application itself and the needs of the design.

Custom resource definitions

As we mentioned in *Chapter 1*, *Introducing the Operator Framework*, most Operators are dependent on **CustomResourceDefinitions** (**CRDs**). CRDs are native Kubernetes resources that allow users to extend the Kubernetes platform with their resource definitions at runtime. They are like plugins to the Kubernetes API. Upon installing a CRD in the cluster, it provides the API server with information on how to handle and validate objects of that custom type.

The most common use case for an Operator to consume CRDs is to provide its configuration object as one. Once the Kubernetes API knows how to handle this CRD, the Operator or user will then create the Operator object from YAML (just like when creating a Pod or Deployment).

Users can then interact solely with the Operator resource, rather than having to manually tweak the Operator's Deployment to adjust its settings. This allows developers to present a curated frontend user experience and abstract away the Kubernetes internals of the application, all in an interface that feels just like any other native Kubernetes resource object.

But Operators can package more than just their configuration in a CRD. Depending on the application that's being managed, an Operator can also install and monitor custom resources that are needed by the application. Because of the many use cases for CRDs, it's not unusual for an application to provide and depend on its custom resource objects. An application Operator should know these resources and be able to install and manage them too.

CRDs are a core aspect of Operator development because, without them, all that is left are the core Kubernetes resources (such as Pods, ReplicaSets, and Deployments) to achieve our goals. While these resources can accomplish many things, they do not have the inherent flexibility to be customized to any specific need. An Operator's custom resource object provides a user interface that seamlessly fits into a cluster with other Kubernetes objects. In addition, the Kubernetes API clients and Operator SDK provide code tools so that you can easily interact with these custom resources, just like they were any other cluster resource.

ServiceAccounts, roles, and RoleBindings (RBAC)

Access policies are often overlooked but can be critical to ensuring the stability and security of your and your users' clusters. Just like an application manages other cluster components, an Operator is going to require certain RBAC policies to complete its job within the necessary access.

Operators require RBAC policies to manage their Operands. Defining an RBAC policy for any application in Kubernetes starts with a **role**. Roles define the types of API objects that an application or user has access to, as well as the verbs that are allowed to be used with those objects. The following is an example of a role:

```
apiVersion: rbac.authorization.k8s.io/v1
kind: Role
metadata:
  namespace: my-operators-namespace
  name: my-operator-role
rules:
- apiGroups:
  - operator.sample.com
  resources:
  - "*"
  verbs:
  - "*"
- apiGroups:
  - ""
  resources:
  - pods
  verbs:
  - get
```

```
    - watch
    - list
```

This sample creates a role that allows an Operator to *get*, *watch*, and *list* the Pods in the specified namespace. It also allows the Operator to access all the resources under the `operator.sample.com` API group. This is included to indicate how an Operator gains access to its CRD. Since there is nothing inherently tying the Operator to the CRD, it still needs RBAC access to that object, just like it would with any other API object. In this case, the CRD would be created under that API group.

Roles can be scoped to a single namespace, or they can be cluster-scoped with **ClusterRoles**. Either way, the role is bound to a **ServiceAccount** with a **RoleBinding** (or **ClusterRoleBinding**). ServiceAccounts are specified on Pods to complete the chain of identifying what RBAC-based access a Pod has within the cluster:

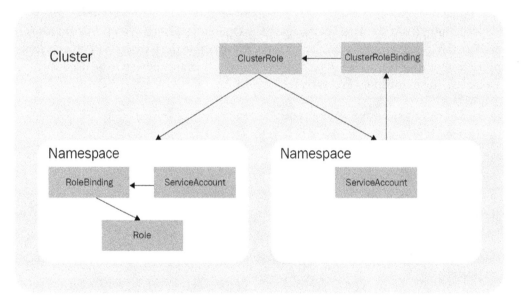

Figure 2.2 – Diagram of cluster versus namespace-scoped roles

The difference between roles and ClusterRoles depends on the breadth of access an Operator will need within a cluster. Specifically, this is mostly based on the namespaces that an Operator, its Operand, and their dependencies are installed in.

Namespaces

All Kubernetes applications run within **namespaces**. A namespace is a logical partition for application resources to reside in that separates components and allows certain degrees of access control. Operators are also deployed within namespaces, but they sometimes require access to other namespaces as well.

The breadth of an Operator's namespace coverage is called its scope. Operators can either be scoped to a single namespace (**namespace-scoped**), or they can be **cluster-scoped**. When an Operator is cluster-scoped, it can observe and interact with resources in multiple namespaces. In contrast, a namespace-scoped operator is installed on a single namespace and manages the resources within that namespace. Which scope a developer chooses for their Operator depends on the resources the Operator will manage.

If the Operator must be able to manage resources that are created in any namespace, then it should be cluster-scoped. This could be an Operator that manages an Operand in a specific separate namespace from the Operator itself. Or, it could simply need to manage Operands that exist in many namespaces (such as a resource provisioner where users request development environments to be deployed).

On the other hand, namespace-scoped Operators have some benefits. Specifically, restricting an Operator to a single namespace allows for easier debugging and failure isolation. It also allows for flexible installation (including installing multiples of the same Operator across multiple namespaces).

The Operator's scope is ultimately defined by its CRD and access roles. Cluster-scoped custom resources are created outside of any specific namespace. This can be good for Operators that will only have one instance in a cluster (for example, Operators that manage core Kubernetes components, such as the API server). The access roles an Operator needs will usually follow the scope of the Operator's CRD.

In summary, the resources we have covered here (Pods, ReplicaSets, Deployments, CRDs, RBAC, and namespaces) are only a few of the possible resources an Operator will depend on. However, they are the most common and likely the first ones you should consider when you're designing your Operator's cluster interactions. In the next section, we'll look at the other side of an Operator's interactions: how it interacts with humans.

Identifying users and maintainers

The other way that an Operator interacts with Kubernetes is through its users. While Operators exist to automate many of the cluster interactions that are required by humans, the organic element in a Kubernetes cluster is still present. Users must still interact with the Operator somehow, and the ways that different users will want and need to engage in that interaction can influence an Operator's design.

Therefore, it is important to identify what type of user your Operator is intended for; there are a few categories. Each category will have different needs and opinions on what their Operator should do. So, by identifying the target audience for your Operator, you can ensure that the Operator has been designed to appeal to the widest user base and fulfill the most use cases.

For most Operators, the type of users that will be interacting with them can be broken into a couple of groups based on the level of access those users have in the cluster and the role those users play in the application at large. These user groups are usually as follows:

- **Cluster administrators**: The admin users of a cluster that are responsible for its maintenance and stability.

- **Cluster users**: Individual users with internal access to a cluster, such as the engineers on a team.

- **End-users/customers**: The external users of a cluster for a public or enterprise product. These users don't directly interact with a cluster, but they use the applications that run on the cluster.

These types of users all have similar motivations and use cases, but also some stark differences. Thinking about each group individually helps highlight these comparisons.

Cluster administrators

Cluster administrators are commonly the consumers of Operators. These are the users that have the most access to and knowledge about a cluster's architecture. As such, they require the most power and flexibility.

Foremost, a cluster administrator's job is to ensure the stability of their cluster. This is why they need the broadest array of tools and controls at their disposal to modify cluster settings as necessary. But, as the saying goes, *with great power comes great responsibility*. In this case, an Operator can provide administrators with great power to run their cluster however they need to. But that power can backfire and harm the cluster if it doesn't work exactly as intended.

So, the developer of an Operator that is going to be used by cluster administrators may consider restricting its capabilities to a defined set that is known to be supportable. This can help them design the Operator so that it doesn't become *too powerful* and expose settings that can inadvertently damage the cluster.

However, if an administrator finds that their cluster is suddenly *on fire* or critically failing, they will need to fix that problem as soon as possible. If this is due to the component that an Operator is managing (or the Operator itself), then the administrator will require direct access to the Operator, Operand, or both. At times like this, a restricted feature set may hinder the recovery time to restore cluster stability.

One benefit of developing Operators for cluster administrators is that they are very knowledgeable about their clusters and have high levels of internal access to them. This provides some trust that allows more powerful functionality to be created. Also, having homogenous consumers limits the different workflows that need to be supported in an Operator.

The following diagram shows the simple layout of an internal cluster administrator interacting with an Operator to manage the Operands in their cluster:

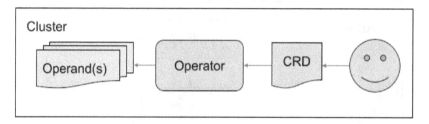

Figure 2.3 – A single cluster administrator directly managing an Operator via its CRD

Ultimately, it is up to the developers of the Operator to determine what to provide for the cluster administrators (if that is the target audience). Depending on the application the Operator is going to manage, it may make sense and be safe and limit the administrator's direct access to the Operand through the Operator. But if there is the potential for an Operand deployment to go haywire and require manual intervention, then emergency accessibility should be provided.

Cluster users

The users of a cluster interact with the cluster but without the access and control of an administrator. For example, these could be developers working on a shared cluster who need certain resources to be deployed and destroyed on-demand. Provisioning development resources involves having users request those resources from the Operator.

Cluster users require less control over underlying cluster components. They also are less likely to be touching parts of a cluster and potentially breaking it. These are two reassurances for developing an Operator that helps limit the scope of features that must be supported by these users. Additionally, these users may have personal access to a cluster administrator if they require more than the Operator provides.

Cluster users may also interact with Operators through internal applications. It is not uncommon for organizations to deploy custom applications within their intranets. Using an internal application like this can provide a more user-friendly frontend for your cluster users. It also allows the cluster users' access to be restricted below the underlying Operator's actual capabilities.

The following diagram shows a setup where multiple internal users are using an Operator to provision resources in the cluster. Some of these users are interacting with the Operator directly, while others must go through an internal application to interact with the Operator (the Operator could even be managing this frontend application). With patterns like this, intra-organizational privilege hierarchies can be defined for working with the internal tools that are provided by Operators:

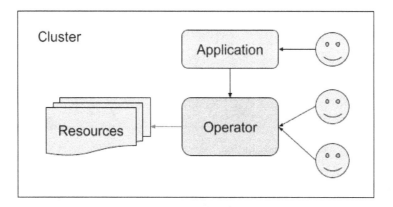

Figure 2.4 – Internal cluster users requesting resources, either directly from the Operator or via an intermediary internal application

The options for exposing an Operator to internal cluster users are some of the most flexible. These users benefit from a level of trust and investment in the cluster's stability that's high enough that the risk of negligent or malicious behavior is mitigated. However, there is still a risk in exposing application management controls to a broader audience. So, you must think about this when you're designing the capabilities of your Operator.

End users and customers

The end users of a product can also benefit from Operators. These users may not even know they are interacting with an Operator since the architectural design of the product is usually not apparent to everyday users. But it is still important to know how the users of a product will expect it to function, especially in this case, where we are designing a critical component of the product.

These end users will likely not be directly interacting with the Operator. This would mean that your users and customers have access to the Kubernetes cluster themselves, which is not ideal for security or usability. The customers of an application benefit from an interactive frontend, whether this is a website or mobile app. But that frontend is just a tool for interacting with a backend, which could be made up of many different components, including Operators.

In this case, your user might be yourself – that is, you (or your organization) will be developing the Operator but also probably developing the frontend applications that depend on it. In situations like this, cross-project collaboration is necessary to elaborate on the needs and expectations of each team. Operators in this kind of scenario will benefit the most from elegant API design that communicates easily with other programs, rather than human users:

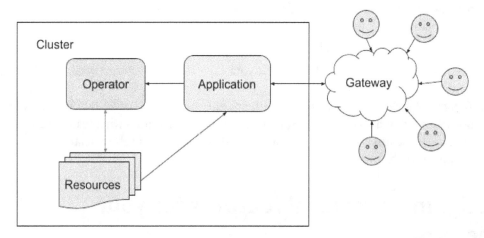

Figure 2.5 – External end users interacting with an Operator via an external-facing application

The end users of an application will usually be completely detached from interacting with an Operator. However, this doesn't mean that an application and its resources can't still be managed by one. In this scenario, the Operator is still an important and helpful tool for maintaining the state of the cluster that end users expect, even if those users don't need to know they are interacting with it.

End users are the final type of users we'll discuss that interact with your Operator in a functional capacity. However, there are other people we must consider when we're designing an Operator: the people who maintain the Operator and its code.

Maintainers

The final type of user who will interact with your Operator does so in a different way than the other three. These are the maintainers of the project, who work on the code to resolve issues and implement new features. The role of a maintainer in any software project is an important one, but in an open source ecosystem such as Kubernetes, there are additional points to consider.

If your Operator's source code is going to be open and accepting contributions from anyone, it will be essential to appoint trusted owners to review code changes. This is true for any open source project, but for an Operator, it is very beneficial that these reviewers are familiar with the core Kubernetes concepts that any Operator depends on.

From an enterprise perspective, investing in a reliable team of engineers to build and maintain an Operator provides a long-term incentive for your maintainers to continue building the Operator's source code. This also creates a level of trust between the maintainers of the Operator and its primary stakeholders. Ongoing maintenance for the Operator code base will be necessary, especially given the changing nature of the Kubernetes platform. Maintainers who are familiar with the Kubernetes community are an important addition to teams that use Operators.

These are just a few of the types of users that your Operator may interact with. This list is by no means exhaustive, but it is intended to provoke thoughts about your user base. Identifying the type of user for an Operator helps narrow down what kind of features an Operator will need. It also provides a starting point for determining what features are necessary. In the next section, we'll look at some ideas for thinking about feature design that benefits users the most.

Designing beneficial features for your operator

Once you've identified the target audience for your Operator, the next step is to define what the Operator will do. Listing the problems that must be solved by the Operator will give you a better understanding of the goals of the project. It will also highlight whether or not those goals provide a tangible, measurable benefit to your users.

Figuring out what kind of functions are truly beneficial is hard to define. The exact definition of "helpful" is different on a case-by-case basis. Some beneficial Operators solve widespread headaches in a novel and intuitive way. Others address more niche problems that may only affect small communities of users, but the impact of eliminating those problems is significant. However, it is a bit easier to describe beneficial functions in terms of what is not useful.

Foremost, useful Operators are not redundant. In specific examples, redundancy can mean a lot of different things. But it is fundamentally the essence of nonduplication that Operator designers should strive for. The most basic example of this is not rewriting an Operator that already exists elsewhere without having a sufficient reason to do so. Researching the proposed area of your Operator can prevent this. Researching the approaches that others have undertaken will also reveal potential pitfalls to avoid in the process.

Operators can also be redundant in terms of Kubernetes concepts. Since an Operator is essentially interacting with the cluster and exposing some of that interaction to users, the Operator is an extension of Kubernetes. Because of this, it can be inadvertently tempting to "reinvent the wheel" when it comes to brainstorming functions for the Operator. This kind of functional redundancy can be related to the next issue we'll discuss, wherein developers try to address nonexistent problems.

The next issue concerns beneficial tasks, which are not hypothetical. The world of software development benefits from many passionate engineers who are eager to fix the problems they run into. This zeal for contribution can extend to fixing potential issues, even if they have not been witnessed in practice. This is certainly a good mentality to have, and in no way should this kind of passion be stifled. There are, of course, many types of issues and bugs that can be identified and should be patched – hopefully before anyone runs into them in production. But occasionally, real solutions are proposed for purely theoretical problems.

These hypothetical use cases may be proposed without any actual evidence that users are seeking such a solution. This can be tricky because sometimes, the use case seems obvious enough that it almost shouldn't necessitate any explicit demand. But when each new feature costs resources to implement and binds the maintainers to support it, paying close attention to silence can reveal unnecessary work.

While there could be many breakthrough ideas for features, it is important to try to validate a feature's proposal with real demand. This type of research can also uncover redundancy in the idea and lead to more effective alternative solutions that already exist.

There are, of course, an infinite number of criteria that could be used to define the usefulness of features. In the context of Kubernetes Operators, however, these are just some initial thoughts to get started with. By thinking about your Operator in terms of its benefits, and by asking "What makes this useful?" you can avoid excessive or unnecessary changes that will require maintenance later down the road.

That foresight will come in handy as your Operator grows and evolves. Like most software, Operators are prone to changes over time, especially given their dependence on the upstream Kubernetes code base. In the next section, we'll learn how to prepare for and navigate these changes as they arrive.

Planning for changes in your Operator

Over any software project's lifespan, changes will be made to the code base. These include bug fixes, refactors, new features, and old features being removed. This is true for Kubernetes, its subprojects, and projects that are built on the platform, such as Operators.

While it's impossible to predict the future of what your Operator will evolve into one day, there are a few ideas that can help you during the design phase to ease transitions and new developments later. These are as follows:

- Start small.
- Iterate effectively.
- Deprecate gracefully.

These decisions have significant benefits in practice when you're maintaining an Operator. However, these are not definitive or strictly specific to Kubernetes Operators. Consider them as general suggestions for developing any software project. But here, we will be examining them in the context of writing an Operator.

Starting small

When planning the original design for an Operator, considering how the Operator may change over time can prevent future challenges. Designing with growth in mind also helps limit the initial scope of the project so that it only covers essential purposes. This allows adequate resources to be allocated to the development of a strong first product that can be efficiently iterated over time.

While it may seem like the best goal for developing an Operator is to start with as many features as possible, in reality, the opposite is often true. After all, one purpose of an Operator is to create a layer of abstraction between its users and the underlying cluster functions. By exposing too many options, you may lose the benefit of automation by requiring your users to understand more of the native Kubernetes cluster functionality. This also makes your users more vulnerable to changes in the Kubernetes platform that shouldn't directly affect them.

When you're deciding whether to add a feature to an Operator or not, think critically about the importance of the feature, as well as the risks and costs of implementing it. This goes back to the topics of avoiding redundancy and addressing hypothetical problems, which we looked at in the previous section. It also relates to the topic of deprecation, which we will cover shortly.

Ultimately, most of the time, it is better to lack features than to ship a complex and confusing Operator. Every new feature has maintenance costs and the potential to introduce bugs later. These are some of the biggest reasons for starting small and building as you listen to feedback from your users.

Iterating effectively

Once you have a minimal set of features, it will eventually be necessary to add improvements. Fortunately, it is much easier to add new things than it is to take old things away.

You should actively seek feedback from users on what kind of features they need. This can be done in many ways, from maintaining an active presence online around your Operator and its related projects to implementing detailed usage metrics in the Operator's code. In addition, having an awareness of the broader Kubernetes project communities can help highlight upcoming changes in the underlying platform that you may want to support in your Operator.

As you add features, continue to monitor their usage to gauge their effectiveness. Keeping the guidelines for the initial design of an Operator, such as its user base and feature tips, in mind will also make sure that each new feature is just as effective as the original set of features you designed when you first built the Operator.

Deprecating gracefully

One type of change that can be incredibly inconvenient for users is deprecation. Many aspects of large software projects inevitably have to be replaced or have their support dropped by official maintainers. With a minimal Operator design, the possibility of your users' experience having to be changed is also minimized.

Starting with a small design and thoughtful, effective iterations after that will help reduce the need to remove features. Unfortunately, it may be inevitable at some point. In this case, it is important to give users plenty of notice so that they have sufficient time to transition to an alternative, if applicable.

The Kubernetes community has defined policies for deprecation, which will be covered in detail *Chapter 8, Preparing for Ongoing Maintenance of Your Operator*. But these guidelines serve as a good template for your deprecation. Kubernetes users will be familiar with the timelines and processes for deprecation, so by remaining consistent with those processes, your users will benefit from a familiarity they already have.

Just as all good things must come to an end, so too must some good features be eventually retired. Fortunately, this is not a foreign concept to the Kubernetes community, which offers a well-established template for deprecation. As a responsible citizen of the open source community, you and your users will benefit from a respectful, considerate approach to removing and changing aspects of your project.

Keeping all these practices in mind during the design stage of an Operator will pay off by limiting the scope of the project. This allows resources to focus on building a stable tool that does not radically change over time. Users and maintainers will both benefit from this stability, as users can seamlessly work with different versions of the Operator without changing their workflows, and maintainers can invest long-term effort into growing the project and passing knowledge about it to others.

Summary

This chapter focused on the different ways in which an Operator can interact with a Kubernetes cluster. Besides the literal technical interactions between an Operator's code base and the cluster's native resources, we also explored some other interactions that are worth considering in an Operator's design. These include an Operator's users and an Operator's lifecycle over time.

Kubernetes clusters comprise many different types of native resources. These are the fundamental building blocks of all the applications that are deployed on Kubernetes. Operators are no different in that sense compared to any other application, so they must be able to natively consume these resources. This chapter focused on breaking down a few Operator resources, including Pods, Deployments, CRDs, and RBAC policies, so that you know how to define how Operators consume them.

How humans interact with an Operator is one of the most important design concepts to consider. Operators are designed to work for humans by automating various tasks, but like any power tool that automates manual labor, they still require human input to operate and produce output for humans. Due to this, we discussed some of the types of users that your Operator can be built for and the unique needs and expectations of each type.

Finally, we covered some good approaches to feature design. To help you set up an Operator for success, we discussed ideas for drafting an initial design that provides tangible benefits to users. We then proposed some more concepts you should keep in mind as your Operator evolves.

In the next chapter, we'll apply these lessons to hands-on Operator design. We will begin by building a sample Operator, starting with designing for its CRD, API, and reconciliation loop. Here, we will begin building an actual Operator, which we will then code and deploy in the remainder of this book.

Part 2: Designing and Developing an Operator

In this section, you will learn how to create your own Operator, from general architectural best practices to specific technical code examples.

This section comprises the following chapters:

- *Chapter 3, Designing an Operator – CRD, API, and Target Reconciliation*
- *Chapter 4, Developing an Operator with the Operator SDK*
- *Chapter 5, Developing an Operator – Advanced Functionality*
- *Chapter 6, Building and Deploying Your Operator*

3

Designing an Operator – CRD, API, and Target Reconciliation

The lessons from the previous chapters have helped us understand the foundational basics of the Operator Framework. In *Chapter 1, Introducing the Operator Framework*, we covered the conceptual pillars of the Operator Framework and the purposes they serve. Then, in *Chapter 2, Understanding How Operators Interact with Kubernetes*, we discussed some general principles of software design in the context of **Kubernetes** and the **Operator Framework**. Together, these chapters have established a baseline understanding of Operators and their development in broad terms. Now, we will be applying this knowledge with examples and begin designing our own Operator.

We'll start by defining a simple problem that our Operator is going to solve. In this case, that will be managing a basic deployment of an application with a single Pod. Over the next few chapters, we will add functionality to this Operator with specific code examples, but before we can start coding our sample Operator, we must first walk through the design process. Building on the generic definitions and steps we've already discussed with a concrete example will provide a context with which to frame the earlier lessons in practical terms.

This process will cover a few different steps in order to lay out the core aspects of our Operator. These will include drawing out the **application programming interfaces (APIs)**, **CustomResourceDefinitions (CRDs)**, and reconciliation logic that will make our Operator work. Along the way, these steps will be related back to the lessons discussed earlier and industry-standard best practices for Operators. We'll break this process into the following steps:

- Describing the problem
- Designing an API and a CRD
- Working with other required resources
- Designing a target reconciliation loop
- Handling upgrades and downgrades
- Using failure reporting

We won't start writing any actual code yet besides some **YAML Ain't Markup Language (YAML)** snippets where applicable. However, we will use some pseudocode to better visualize how our actual code will work once we initialize the project with the Operator **software development kit (SDK)** in *Chapter 4, Developing an Operator with the Operator SDK.*

Describing the problem

Many software projects can be defined with a user story in the following format: *As a [user] I want to [action], so that [reason].* We'll also do that here, as follows:

> *As a cluster administrator, I want to use an Operator to manage my nginx application so that its health and monitoring are automatically managed for me.*

For our use case (designing an Operator), we don't care about the specific application right now. For that reason, our *application* is just going to be a basic nginx sample Pod. We will assume that this represents any single-Pod application with basic **input/output (I/O)** needs. While this may seem too abstract, the focus will be on building an Operator around the application.

The first thing we have identified from this user story is that we will be building the Operator for cluster administrators. From the previous chapter, we know this means that the Operator's users will have a better understanding of the cluster architecture than most end users and that they will need higher levels of direct control over the Operand. We can also assume that most cluster administrators will be comfortable interacting directly with the Operator rather than through an intermediary frontend application.

The second part of this user story identifies the functional objective of the Operator. Specifically, this Operator is going to manage the Deployment of a single-Pod application. In this case, *manage* is a vague term that we will assume to mean create and maintain the required Kubernetes resources to run the application. These resources will be, at minimum, a Deployment. We will need to expose some of the options of this Deployment through the Operator, such as the container port for nginx.

Finally, the user story provides our motivation for running the Operator. The cluster administrators want to have the application's *health and monitoring* managed by the Operator. Application health can mean a lot of different things, but generally, this comes down to maintaining high uptime for the application and recovering from any crashes, if possible. Monitoring the application can also be done in a number of ways, usually in the form of metrics.

So, from all of the preceding information, we have identified that we want a very basic Operator that can do the following:

- Deploy an application
- Keep the application running if it fails
- Report on the health status of the application

These are some of the simplest functions that an Operator can serve. In later chapters, we'll build on these requests a bit more. But in the interest of starting with a solid foundation upon which to iterate later, this will be our **minimum viable product** (**MVP**). Henceforward, therefore, this is the basic Operator design we will be referencing when referring to our examples.

Based on these criteria, we can try to define our Operator in terms of the Capability Model covered in *Chapter 1, Introducing the Operator Framework* (recall that the Capability Model defines five levels of Operator functionality, from *Basic Install* to *Auto Pilot*). We know that the Operator will be able to install the Operand, as well as manage any additional required resources. We would like it to be able to report on the status of the Operand as well and provide configuration through its CRD. These are all the criteria for a Level I Operator. In addition, it would be great if our Operator could handle upgrades to qualify it as a Level II Operator.

This is a good start for the initial Operator design. With a full picture of the problem we are trying to solve, we can now begin to brainstorm how we will solve it. To do that, we can start by designing how our Operator will be represented in the cluster API.

Designing an API and a CRD

As we covered in *Chapter 1, Introducing the Operator Framework*, and *Chapter 2, Understanding How Operators Interact with Kubernetes*, the use of a CRD is a defining characteristic of Operators to create an object for users to interact with. This object creates an interface for controlling the Operator. In this way, the **Custom Resource** (**CR**) object is a window into the Operator's main functions.

As with any good window, the Operator's CRD must be built well. It must be clear enough to expose the details of the Operator while being secure enough to keep out harsh weather and burglars, and as with a window, the CRD's design should follow local building codes to ensure that it is built up to the expected standards of the environment. In our case, those building codes are the Kubernetes API conventions.

Following the Kubernetes API design conventions

Even though a CRD is a custom object that can be created by anyone, there are still best practices to keep in mind. This is because the CRD exists within the Kubernetes API, which defines its conventions so that there are certain expectations when interacting with the API. These conventions are documented in the Kubernetes community at `https://github.com/kubernetes/community/blob/master/contributors/devel/sig-architecture/api-conventions.md`. However, this documentation is extensive and covers the requirements for all kinds of API objects, not just Operator CRDs. There are, however, a few key elements that are relevant to our purpose, as outlined here (Note: some of these fields will be discussed in more detail later in the chapter):

- All API objects must have two fields: `kind` and `apiVersion`. These fields make it possible for Kubernetes API clients to decode the object. `Kind` represents the name of the object type—for example, `MyOperator`—and `apiVersion` is, aptly, the version of the API for that object. For example, your Operator may ship API versions `v1alpha1`, `v1beta1`, and `v1`.

- API objects should also have the following fields (though they are not required):

  - `resourceVersion` and `generation`, both of which help to track changes to an object. However, these fields serve different purposes. The `resourceVersion` field is an internal reference that is incremented every time an object is modified, which serves to help with concurrency control. For example, when trying to update an Operand Deployment, you will make two client calls: `Get()` and `Update()`. When calling `Update()`, the API can detect if the object's `resourceVersion` field has changed on the server (which would indicate that another controller has modified the object before we updated it) and reject the update. In contrast, `generation` serves to keep track of the relevant updates to an object. For example, a Deployment that has recently rolled out a new version would have its `generation` field updated. These values can be used to reference older generations or ensure that new ones are of the expected generation number (that is, *current+1*).

  - `creationTimestamp` and `deletionTimestamp`, which serve as helpful reference points for the age of an object. With `creationTimestamp`, for example, you can easily reference the age of an Operator's Deployment based on when its CRD was created. Similarly, `deletionTimestamp` serves to indicate that a deletion request has been sent to the API server for that object.

- `labels` and `annotations`, which serve similar purposes but are semantically different. Applying `labels` to an object serves to organize objects by criteria that are easily filtered through the API. On the other hand, `annotations` exposes metadata about the object.

- API objects should have `spec` and `status` fields. We will cover `status` in more detail later in the chapter (under *Using failure reporting*), but for now, there are some conventions around it to keep in mind, as outlined here:

  - Conditions reported in an object's `status` field should be clearly self-explanatory without the need for additional context to understand them.

  - Conditions should respect the same API compatibility rules as any other field. In order to maintain backward compatibility, condition definitions should not change once defined.

  - Conditions can report either `True` or `False` as their normal operating state. There is no set guideline for which one should be the standard mode; it is up to the developer to consider readability in the definition of the condition. For example, a `Ready=true` condition can have the same meaning as one called `NotReady=false`, but the former is much easier to understand.

  - Conditions should represent the current known state of the cluster, rather than reporting transitions between states. As we will cover in the *Designing a target reconciliation loop* section, many Kubernetes controllers are written with a level-triggered design (meaning they operate based on the current state of the cluster rather than incoming events alone). So, an Operator reporting conditions based on its current state helps to maintain this mutual design assumption of being able to build the current state of the cluster in memory at any time. However, for long transitionary phases, the `Unknown` condition can be used if necessary.

- Sub-objects within an API object should be represented as lists, rather than maps; for example, our nginx deployment may need several different named ports to be specified through the Operator CRD. This means that they should be represented as a list, with each item in the list having fields for `name` and `port`, as opposed to a map where the key for each entry is the `name` of the port.

- Optional fields should be implemented as pointer values in order to easily distinguish between zero and unset values.

- The convention for fields with units is to include the units in the field name—for example, `restartTimeoutSeconds`.

These are just some of the many API conventions, but they are important to know as you design your Operator's CRD and API. Adhering to the guidelines of API design ensures that other components in the Kubernetes ecosystem (including the platform itself) can make appropriate assumptions about your Operator. With these guidelines in mind, we can move on to the next step of designing our own CRD schema.

Understanding a CRD schema

We have already discussed CRDs and their importance to Operators, but up until this point, we haven't looked in depth at how a CRD is composed. Now that we know the problem our example Operator is going to solve, we can begin looking at the options we want to expose through its CRD and get an idea of what those will look like to users.

First, it is best to see an example CRD and examine each section to understand its purpose, as follows:

```yaml
apiVersion: apiextensions.k8s.io/v1
kind: CustomResourceDefinition
metadata:
  name: myoperator.operator.example.com
spec:
  group: operator.example.com
  names:
    kind: MyOperator
    listKind: MyOperatorList
    plural: myoperators
    singular: myoperator
  scope: Namespaced
  versions:
    - name: v1alpha1
      schema:
        openAPIV3Schema:
          . . .
      served: true
      storage: true
      subresources:
        status: {}
status:
  acceptedNames:
```

```
    kind: ""
    plural: ""
conditions: []
storedVersions: []
```

The first two fields, `apiVersion` and `kind`, define that this is a CRD. Even though we are trying to define a blueprint of our own custom objects, that blueprint must exist within a `CustomResourceDefinition` object first. From this, the API server will know how to parse the CRD data to create instances of our CR.

Next, the `metadata.Name` field defines the name of our CRD (not the custom object created from the CRD). To be more specific, this is the name of the blueprint, not the objects created from the blueprint. For example, we could retrieve this CRD design using `kubectl get crd/myoperator.operator.example.com`.

Within `spec` is where the CRD begins to actually define the CR objects we want to create. The `group` defines a custom API group to which new objects will belong. Using a unique name here helps to prevent collisions with other objects and APIs in the cluster.

The `names` section defines different ways in which our objects can be referenced. Here, only `kind` and `plural` are required (as the others can be inferred from these two). Just as any other type of object in the cluster is accessible via its `kind` or `plural` form (for example, `kubectl get pods`), our CRs will be accessible the same way with commands such as `kubectl edit myoperator/foo`. Even though most Operators will (and should) only have one CR object in the cluster, these fields are still required.

Next, `scope` defines custom objects as namespace- or cluster-scoped. The differences between these two were covered in detail in *Chapter 1*, *Introducing the Operator Framework*. The available options for this field are `Cluster` and `Namespaced`.

`Versions` provides a list of the different API versions that will be available for our CR. As your Operator evolves over time and new features are added or removed, you will need to introduce new versions of the Operator's CR. For backward compatibility and support, you should continue to ship older versions of the resource to provide users a transitory period after which the version can be safely deprecated. This is why this field provides a list of versions. The API server is aware of each version and can operate effectively on an object that is created and used in any available version in this list.

Each version in the list contains schematic information about the object itself that uniquely identifies the structure of that version in `openAPIV3Schema`. In this example, the `openAPIV3Schema` section has been intentionally omitted. We have done that because this section is usually very long and complex. However, in recent versions of Kubernetes, this section is required in order to provide a **structural schema** for the CRD.

A structural schema is an object schema that is based on **OpenAPI version 3 (V3) validation**. OpenAPI defines validation rules for each field that can be used to validate field data when objects are created or updated. These validation rules include the type of data for the field, as well as other information such as allowed string patterns and enumerated values. The structural schema requirement for CRDs ensures consistent, reliably stored representations of the objects.

Due to the complex nature of OpenAPI validation schemas, it is not recommended to write them by hand. Instead, the use of generated tools such as **Kubebuilder** (which is used by the Operator SDK) is recommended. Extensive and flexible validation rules can be defined directly on the Go types for CRDs using the various Kubebuilder markers, which are available for full reference at `https://book.kubebuilder.io/ reference/generating-crd.html`.

The next sections of the individual version definitions are `served` and `storage`, which set whether this version is served via REST APIs and if this is the version that should be used as the storage representation. Only one version can be set as the storage version for a CRD.

The final sections, `subresources` and `status`, are related because they define a `status` field that will be used to report information on the current state of the Operator. We will cover that field and its uses in more detail under *Using failure reporting*.

Now that we have explored the structure of a CRD and have an idea of what one should look like, we can design one for our example nginx Operator.

Example Operator CRD

From the preceding problem statement, we know that our Operator is initially going to simply deploy an instance of nginx in the cluster. We also now know that our Operator's CRD will provide a `spec` field with various options to control the Operand Deployment. But what kind of settings should we expose? Our Operand is fairly simple, so let's start with a few basic options that we are defining to configure a simple nginx Pod, as follows:

- `port`—This will be the port number that we want to expose the nginx Pod on within the cluster. Since nginx is a web server, this will allow us to modify the accessible port without having to directly touch the nginx Pod, because the Operator will handle safely changing it for us.

- `replicas`—This is a bit redundant because the number of replicas for any Kubernetes Deployment can be adjusted through the Deployment itself. But in the interest of abstracting control of the Operand away into a **user interface** (**UI**) behind the management of an Operator, we will provide this option. This way, an administrator (or other application) can scale the Operand with the added handling and reporting of our Operator.

- `forceRedeploy`—This field is interesting because it will effectively be a **no-operation** (**no-op**) against the Operand in terms of how it functions. However, including a field in the Operator CRD that can be set to any arbitrary value allows us a way to instruct the Operator to trigger a new rollout of the Operand without having to modify any actual settings. Having that functionality is useful for stuck Deployments where manual intervention can resolve the issue.

This works because the Operator watches for changes to relevant resources in the cluster, one of which being its own CRD (more on this in the *Designing a target reconciliation loop* section). This watch is necessary so that the Operator knows when to update an Operand. So, including a no-op field can be enough for the Operator to know to redeploy the Operand without needing to make any actual changes.

These three settings together will make the basis of our Operator's CRD `spec`. With this, we know that as an object in the cluster the CR object will look something like this:

```
apiVersion: v1alpha1
kind: NginxOperator
metadata:
  name: instance
spec:
  port: 80
  replicas: 1
status:
  . . .
```

Note that this is the CR itself, not the CRD, as shown in an example earlier. We are using a generic `name: instance` value here because we will probably only have one instance of the Operator running in a namespace at a time. We also haven't included the `forceRedeploy` field here because that will be optional.

This object could be retrieved with the `kubectl get -o yaml nginxoperator/instance` command if we define our CRD right. Thankfully, the Operator SDK and Kubebuilder will help us generate that.

Working with other required resources

Besides the CRD, our Operator will be responsible for managing a number of other cluster resources as well. Right now, this is the nginx Deployment that will be created as our Operand, as well as a ServiceAccount, Role, and RoleBinding for the Operator. What we need to understand is how the Operator will know the definition of those resources.

Somewhere, the resources need to be written as Kubernetes cluster objects. Just as you would create a Deployment by hand (for example, with `kubectl create -f`), the definitions of necessary resources can be packaged with the Operator code in a couple of different ways. This can be done easily with templates if you are creating your Operator with Helm or Ansible, but for Operators written in **Go**, we need to consider our options.

One way to package these resources so that the Operator can create them is by defining them directly in the Operator's code. All Kubernetes objects are based on corresponding Go type definitions, so we have the ability to create Deployments (or any resource, for that matter) directly in the Operator by declaring the resources as variables. Here's an example of this:

```go
...
import appsv1 "k8s.io/api/apps/v1"
...
nginxDeployment := &appsv1.Deployment{
  TypeMeta: metav1.TypeMeta{
    Kind: "Deployment",
    apiVersion: "apps/v1",
  },
  ObjectMeta: metav1.ObjectMeta{
    Name: "nginx-deploy",
    Namespace: "nginx-ns",
  },
  Spec: appsv1.DeploymentSpec{
    Replicas: 1
    Selector: &metav1.LabelSelector{
      MatchLabels: map[string]string{"app":"nginx"},
    },
    Template: v1.PodTemplateSpec{
      Spec: v1.PodSpec{
        ObjectMeta: metav1.ObjectMeta{
          Name: "nginx-pod",
```

```go
            Namespace: "nginx-ns",
            Labels: map[string]string{"app":"nginx"},
        },
        Containers: []v1.Container{
            {
                Name: "nginx",
                Image: "nginx:latest",
                Ports: []v1.ContainerPort{{ContainerPort:
int32(80)}},
            },
        },
    },
  },
 },
}
```

The convenience of defining objects in code in this way is helpful for development. This approach provides a transparent definition that is clearly available and immediately usable by the Kubernetes API clients. However, there are some drawbacks to this. First, it is not very human-readable. Users will be familiar with interacting with Kubernetes objects represented as YAML or **JavaScript Object Notation** (**JSON**), wherein the type definitions for each field are not present. This information is unnecessary and superfluous for most users. So, any users who are interested in seeing the resource definitions clearly or modifying them may find themselves lost deep in the Operator's code.

Fortunately, there is an alternative to defining resources as Go types directly. There is a helpful package called go-bindata (available at github.com/go-bindata/go-bindata) that compiles declarative YAML files into your Go binary so that they can be accessible by code. Newer versions of Go (1.16+) also now include the go:embed compiler directive to do this without an external tool such as go-bindata. So, we can simplify the preceding Deployment definition like so:

```yaml
kind: Deployment
apiVersion: apps/v1
metadata:
  name: nginx-deploy
  namespace: nginx-ns
spec:
  replicas: 1
```

```
selector:
  matchLabels:
      app: nginx
template:
  metadata:
    labels:
        app: nginx
  spec:
    containers:
    - name: nginx
        image: nginx:latest
        ports:
        - containerPort: 80
```

This is much more readable for the average user. It is also easily maintained, and you can provide different versions of various resources within named directories in your Operator's code base. This is good for code organization, and also simplifies your options for **continuous integration (CI)** checks against the validity of these type definitions.

We will cover how to use `go-bindata` and `go:embed` in more detail in *Chapter 4, Developing an Operator with the Operator SDK*, but for now, we know how we can package our additional resources to be available in the Operator. This is a key design consideration that benefits our users and maintainers.

Designing a target reconciliation loop

Now that we have defined our Operator's UI by designing a CRD to represent it in the cluster and itemized the Operand resources that it will manage, we can move on to the core logic of the Operator. This logic is nestled within the main reconciliation loop.

As described in earlier chapters, Operators function on the premise of reconciling the current state of the cluster with the desired state set by users. They do this by periodically checking what that current state is. These checks are usually triggered by certain events that are related to the Operand. For example, an Operator will monitor the Pods in its target Operand namespace and react to the creation or deletion of a Pod. It is up to the Operator developers to define which events are of interest to the Operator.

Level- versus edge-based event triggering

When an event triggers the Operator's reconciliation loop, the logic does not receive the context of the whole event. Rather, the Operator must re-evaluate the entire state of the cluster to perform its reconciliation logic. This is known as **level-based triggering**. The alternative to this kind of design is **edge-based triggering**. In an edge-based system, the Operator logic would function only on the event itself.

The trade-off between these two system designs is in efficiency for reliability. Edge-based systems are much more efficient because they do not need to re-evaluate the entire state and can only act on the relevant information. However, an edge-based design can suffer from inconsistent and unreliable data—for example, if events are lost.

Level-based systems, on the other hand, are always aware of the entire state of the system. This makes them more suitable for large-scale distributed systems such as Kubernetes clusters. While these terms originally stem from concepts related to electronic circuits, they also relate well to software design in context. More information is available at `https://venkateshabbarapu.blogspot.com/2013/03/edge-triggered-vs-level-triggered.html`.

Understanding the difference between these design choices allows us to think about how the reconciliation logic will function. By going with a level-based triggering approach, we can be sure that the Operator will not lose any information or miss any events, as the cluster state representation in its memory will always eventually catch up to reality. However, we must consider the requirements for implementing a level-based design. Specifically, the Operator must have the information necessary to build the entire relevant cluster state in memory each time an event triggers reconciliation.

Designing reconcile logic

The reconcile loop is the core function of the Operator. This is the function that is called when the Operator receives an event, and it's where the main logic of the Operator is written. Additionally, this loop should ideally be designed to manage one CRD, rather than overload a single control loop with multiple responsibilities.

When using the Operator SDK to scaffold an Operator project, the reconciliation loop will have a function signature like this:

```
func (r *Controller) Reconcile(ctx context.Context, req ctrl.
Request) (ctrl.Result, error)
```

This function is a method of a `Controller` object (which can be any name; we use `Controller` in this example, but it could just as easily be `FooOperator`). This object will be instantiated during the startup of the Operator. It then takes two parameters: `context.Context` and `ctrl.Request`. Finally, it returns a `ctrl.Result` parameter and, if applicable, an `error` parameter, We will go into more detail about these types and their specific roles in *Chapter 4, Developing an Operator with the Operator SDK*, but for now, understand that the core reconciliation loop of an Operator is built upon very little information about the event that triggered reconciliation. Note that the Operator's CRD and information about the cluster state are not passed to this loop; nor is anything else.

Because the Operator is level-driven, the `Reconcile` function should instead build the status of the cluster itself. In pseudocode, this often looks something like this:

```
func Reconcile:
    // Get the Operator's CRD, if it doesn't exist then return
    // an error so the user knows to create it
    operatorCrd, error = getMyCRD()
    if error != nil {
      return error
    }
    // Get the related resources for the Operator (ie, the
    // Operand's Deployment). If they don't exist, create them
    resources, error = getRelatedResources()
    if error == ResourcesNotFound {
      createRelatedResources()
    }
    // Check that the related resources relevant values match
    // what is set in the Operator's CRD. If they don't match,
    // update the resource with the specified values.
    if resources.Spec != operatorCrd.Spec {
      updateRelatedResources(operatorCrd.Spec)
    }
```

This will be the basic layout for our Operator's reconciliation loop as well. The general process, broken into steps, is this:

1. First, check for an existing Operator CRD object. As we know, the Operator CRD contains the configuration settings for how the Operator should function. It's considered a best practice that Operators should not manage their own CRD, so if there isn't one on the cluster, then immediately return with an error. This error will show up in the Operator's Pod logs and indicate to the user that they should create a CRD object.

2. Second, check for the existence of relevant resources in the cluster. For our current use case, that will be the Operand Deployment. If the Deployment does not exist, then it is the Operator's job to create it.

3. Finally, if relevant resources already existed on the cluster, then check that they are configured in accordance with the settings in the Operator CRD. If not, then update the resources in the cluster with the intended values. While we could just update every time (because we know the desired state without having to look at the current state), it's a good practice to check for differences first. This helps minimize excessive API calls over updating indiscriminately. Making unnecessary updates also increases the chance of an update hot loop, where the Operator's updates to the resources create events that trigger the reconciliation loop that handles that object.

These three steps rely heavily on access to the Kubernetes API via the standard API clients. The Operator SDK provides functions that help make it easy to instantiate these clients and pass them to the Operator's control loop.

Handling upgrades and downgrades

As Operator developers, we are concerned with the versioning of two primary applications: the Operand and the Operator itself. Seamless upgrades are also the core feature of a Level II Operator, which we have decided is our goal for the initial Operator design. For that reason, we must ensure that our Operator can handle upgrades for both itself and the nginx Operand. For our use case, upgrading the Operand is fairly straightforward. We can simply pull the new image tag and update the Operand Deployment. However, if the Operand changed significantly, then the Operator may also need to be updated in order to properly manage the new Operand version.

Operator upgrades arise when changes to the Operator code, API, or both need to be shipped to users. **Operator Lifecycle Manager** (**OLM**) makes upgrading Operators with newer released versions easy from a user standpoint. The Operator's **ClusterServiceVersion** (**CSV**) allows developers to define specific upgrade paths for maintainers to provide specific information about new versions that replace older versions. This will be covered in more detail in *Chapter 7, Installing and Running Operators with the Operator Lifecycle Manager*, when we actually write a CSV for our Operator.

There may also be a scenario where the Operator's CRD changes in incompatible ways (for example, deprecation of an old field). In this scenario, your Operator's API version should be increased (for example, from `v1alpha1` to `v1alpha2` or `v1beta1`). The new version should also be shipped with the existing version's CRD. This is the reason why the CRD's `versions` field is a list of version definitions, and it allows users the ability to transition from one version to the next thanks to simultaneous support of both.

Recall, however, that out of this list of versions there may only be one that is the designated storage version. It is also excessive to ship every previous API version forever (eventually, older versions will need to be completely removed after an appropriate deprecation timeline has passed). When it is time to permanently remove support for deprecated API versions, the storage version may also need to be updated as well. This can cause issues for users who still have the old version of the Operator CRD installed as the storage version in their cluster. The `kube-storage-version-migrator` tool (`https://github.com/kubernetes-sigs/kube-storage-version-migrator`) helps with this by providing a migration process for existing objects in the cluster. The storage version can be migrated with a `Migration` object, such as this:

```
apiVersion: migration.k8s.io/v1alpha1
kind: StorageVersionMigration
metadata:
  name: nginx-operator-storage-version-migration
spec:
  resource:
    group: operator.example.com
    resource: nginxoperators
    version: v1alpha2
```

When this object is created, `kube-storage-version-migrator` will see it and update any existing objects that are stored in the cluster to the specified version. This only needs to be done once, and it can even be automated by packaging this object as an additional resource in the Operator. Future versions of Kubernetes will automate this process fully (see *KEP-2855*, `https://github.com/kubernetes/enhancements/pull/2856`).

Preparing for successful version transitions early on will pay off with future maintenance of your Operator. However, not everything can always go smoothly, and it's impossible to prepare for every possible scenario. This is why it is important for an Operator to have sufficient error reporting and handling as well.

Using failure reporting

When it comes to failures, there are two things we need to worry about: failures in the Operand, and failures in the Operator itself. Sometimes, these two instances may even be related (for example, the Operand is failing in an unexpected way that the Operator does not know how to resolve). When any errors occur, it's an important job of the Operator to report those errors to the user.

When an error happens during the Operator's reconciliation loop, the Operator must decide what to do next. In implementations with the Operator SDK, the reconcile logic is able to identify when an error has occurred and attempt the loop again. If the error continues to prevent the reconciliation from succeeding, the loop can exponentially back off and wait longer between each attempt in the hope that whichever condition is causing the failure will be resolved. However, when an Operator reaches this state, the error should still be exposed to the user in some way.

Error reporting can be easily done in a few ways. The main methods for reporting failures are logging, status updates, and events. Each approach offers different advantages, but a sophisticated Operator design will utilize all three in elegant harmony.

Reporting errors with logging

The simplest way to report any error is with basic logging. This is the way that most software projects report information to the user, not just Kubernetes Operators. That's because logged output is fairly easy to implement and intuitive to follow for most users. This reasoning is especially true considering the availability of logging libraries such as **Klog**, which help standardize the structure of logs specifically for Kubernetes applications. When a Pod is running, the logs are easily retrieved with commands such as `kubectl logs pod/my-pod`. However, there are some downsides to relying on just logging for significant errors.

First, Kubernetes Pod logs are not persistent. When a Pod crashes and exits, its logs are only available until the failed Pod is cleaned up by the cluster's **garbage collection** (**GC**) processes. This makes debugging a failure particularly difficult as the user is in a race against time. Additionally, if an Operator is working diligently to fix the issue, then the user is also racing against their own automation system, which is supposed to help them, not hinder them.

Second, logs can be a lot of information to parse. Besides just the relevant Operator logs you may write yourself, your Operator will be built on many libraries and dependencies that inject their own information into the logged output. This can create a cumbersome mess of logs that require work to sort through. While there are, of course, tools such as grep that make it relatively easy to search through lots of text, your users may not always know exactly which text to search for in the first place. This can create serious delays when debugging an issue.

Logs are helpful for tracing procedural steps in an Operator or for low-level debugging. However, they are not great at bringing failures to the immediate attention of users. Pod logs do not last long, and they are often drowned out by irrelevant logs. In addition, logs themselves usually do not provide much human-readable context for debugging. This is why important failures that require attention are better handled by status updates and events.

Reporting errors with status updates

As mentioned earlier in the chapter when discussing Kubernetes API conventions and CRD design, an Operator CRD should provide two important fields: spec and status. While spec represents the desired state of the cluster and accepts input from the user, status serves to report the current state of the cluster and should only be updated as a form of output.

By utilizing the status field to report the health of your Operator and its Operand, you can easily highlight important state information in a readable format. This format is based on the condition type, which is provided by the Kubernetes API machinery.

A condition reports its name along with a Boolean value indicating whether the condition is currently present. For example, an Operator could report the OperandReady=false condition to show that the Operand is not healthy. There is also a field within the condition called Reason, which allows developers to provide a more readable explanation of the current status. As of Kubernetes 1.23, the Condition Type field has a maximum length of 316 characters, and its Reason field can be up to 1,024 characters.

The Kubernetes API clients provide functions to report conditions easily, such as `metav1.SetStatusCondition(conditions *[]metav1.Condition, newCondition metav1.Condition)`. These functions (and the `Condition` type itself) exist under the `k8s.io/apimachinery/pkg/apis/meta/v1` package.

In an Operator's CRD `status` field, the conditions look similar to this:

```
status:
  conditions:
    - type: Ready
      status: "True"
      lastProbeTime: null
      lastTransitionTime: 2018-01-01T00:00:00Z
```

For our nginx deployment Operator, we'll start by reporting a condition that is simply called `Ready`. We'll set the status of this condition to `True` on the successful startup of the Operator, and change it to `False` in the event that the Operator fails a reconciliation loop (along with a `Reason` field explaining the failure in more detail). We may end up finding more Conditions that will make sense to add later, but given the initial simplicity of the Operator, this should be sufficient.

Using conditions helps show the current state of the Operator and its managed resources, but these only show up in the `status` section of the Operator's CRD. However, we can combine them with events to make the error reporting available throughout the cluster.

Reporting errors with events

Kubernetes events are a native API object, just like Pods or any other object in the cluster. Events are aggregated and show up when using `kubectl describe` to describe a Pod. They can also be monitored and filtered by themselves with `kubectl get events`. Their availability within the Kubernetes API makes them understandable by other applications as well, such as alerting systems.

An example of listing a Pod's events is shown here, where we see five different events:

1. A `Warning` event that has occurred three times, showing that the Pod failed to be scheduled.

2. A `Normal` event once the Pod was successfully scheduled.

3. Three more `Normal` events as the Pod's container images were pulled, created, and successfully started.

You can see these events in the following code snippet:

```
$ kubectl describe pod/coredns-558bd4d5db-6mqc2 -n kube-system
...
Events:
  Type        Reason            Age                         From
Message
  ----        ------            ----                        ----
-------
  Warning  FailedScheduling  6m36s (x3 over 6m52s)  default-
scheduler            0/1 nodes are available: 1 node(s) had
taint {node.kubernetes.io/not-ready: }, that the pod didn't
tolerate.
  Normal   Scheduled         6m31s                      default-
scheduler            Successfully assigned kube-system/coredns-
558bd4d5db-6mqc2 to kind-control-plane
  Normal   Pulled            6m30s                      kubelet,
kind-control-plane  Container image "k8s.gcr.io/coredns/
coredns:v1.8.0" already present on machine
  Normal   Created           6m29s                      kubelet,
kind-control-plane  Created container coredns
  Normal   Started           6m29s                      kubelet,
kind-control-plane  Started container coredns
```

Events can relay more information than conditions thanks to a much more complex object design. While events include a Reason and a Message field (analogous to conditions' Type and Reason fields, respectively), they also include information such as Count (which shows the number of times this event has occurred), ReportingController (which shows the originating controller of the event), and Type (which can be used to filter events of different severity levels).

The Type field can currently be used to categorize cluster events as Normal or Warning. This means that, similar to how a condition can report a successful state, events can also be used to show that certain functions completed successfully (such as startup or upgrades).

For a Pod to report events to the cluster, the code needs to implement an EventRecorder object. This object should be passed throughout the controller and broadcasts events to the cluster. The Operator SDK and Kubernetes clients provide boilerplate code to set this up properly.

Besides reporting events, your Operator will also react to events in the cluster. This goes back to the essential foundation of an Operator's event-triggered design. There are code patterns to design which types of events the Operator is interested in reacting to, wherein you can add logic to filter out specific events. This will be covered in detail in later chapters.

As you can see, a sophisticated error-reporting system utilizes logs, status, and events to provide a full picture of the state of the application. Each method provides its own benefits, and together they weave a beautiful tapestry of debuggability that helps administrators track down failures and resolve issues.

Summary

This chapter outlined the details of an Operator we would like to build. Beginning with a description of the problem (in this case, a simple Operator to manage an nginx Pod) gave a solid foundation of the solutions that are available to work with. This step even provided enough information to set a goal for the capability level of this Operator (*Level II – Seamless Upgrades*).

The next step was outlining what the Operator CRD will look like. To do this, we first noted some relevant conventions in the Kubernetes API that are helpful to ensure the Operator conforms to expected standards for Kubernetes objects. We then broke down the structure of a CRD and explained how each section relates to the corresponding CR object. Finally, we drafted an example of what the Operator's CR will look like in the cluster to get a concrete idea of the expectation from users.

After designing the CRD, we considered our options for managing additional resources as well. For an Operator written in Go, it makes sense to package additional resources (such as RoleBinding and ServiceAccount definitions) as YAML files. These files can be compiled into the Operator binary with `go-bindata` and `go:embed`.

The next step in the design is the target reconciliation loop. This comprises the core logic of the Operator and is what makes the Operator a useful, functional application. This process began with understanding the difference between level- and edge-triggered event processing and why it is better for Operators to be level-based. We then discussed the basic steps of an Operator's reconcile loop.

The last two sections discussed the topics of upgrades, downgrades, and error reporting. With upgrades and downgrades, we covered the use cases for shipping and supporting multiple API versions simultaneously, as well as the need to occasionally migrate storage versions in existing installations. The section about error reporting focused on the three main ways that applications can expose health information to users: logging, status conditions, and events.

In the next chapter, we will take everything we have decided on as our initial design and compose it into actual code. This will involve initializing a project with the Operator SDK, generating an API that will become the Operator's CRD, and coding target reconciliation logic. Essentially, we will apply the knowledge from this chapter to a hands-on exercise in Operator development.

4

Developing an Operator with the Operator SDK

With a completed design outline for an **Operator**, it is now possible to begin the actual work of developing it. This means writing and compiling code that can be deployed onto an actual running **Kubernetes cluster**. For this chapter, the **Operator SDK** will be used to initialize the scaffolding of a boilerplate Operator project. From there, the technical steps to develop the rest of a basic Operator will be demonstrated as a tutorial. This guide will follow the Operator design already planned in *Chapter 3, Designing an Operator – CRD, API, and Target Reconciliation*, which focused on developing a *Level II* Operator to deploy and upgrade a simple **Nginx Pod**.

As a tutorial, this chapter will follow the process for building an Operator from scratch with **Go**. Beginning with the initialization of boilerplate project code, the guide will then follow through the steps of defining the Operator API and generating the corresponding **CustomResourceDefinition** (**CRD**). Then, we will see how to implement simple reconciliation logic that makes up the core functionality of the Operator. Finally, some basic troubleshooting and common issues will be addressed. The steps for developing an Operator with Operator SDK will be broken down into the following sections:

- Setting up your project

- Defining an API

- Adding resource manifests

- Writing a control loop

- Troubleshooting

These sections follow roughly the design pattern that is recommended in the official Operator SDK Go documentation (`https://sdk.operatorframework.io/docs/building-operators/golang/`), which is why we have chosen to follow this approach. At the end of this chapter, we will have an Operator that covers the Level II functionality described in the design that was outlined in *Chapter 3*, *Designing an Operator – CRD, API, and Target Reconciliation*. This functionality includes basic deployment of the Operand (in this case, Nginx) and seamless upgrades of the Operator and Operand. In later chapters, this guide will be built upon as a foundation for adding more complex functionality that graduates this sample Operator from lower to higher levels along the Capability Model.

Technical requirements

The guided steps in this chapter will require the following technical prerequisites to follow along:

- `go` version 1.16+

- An `operator-sdk` binary installed locally

The `operator-sdk` binary can be installed either directly from a release, with Homebrew (for macOS), or compiled from GitHub from `https://github.com/operator-framework/operator-sdk`. If choosing to install the Operator SDK from GitHub, `git` will also be required; however, it is recommended to use `git` anyway to take advantage of version control for the project.

The Code in Action video for this chapter can be viewed at: `https://bit.ly/3N7yMDY`

Setting up your project

The first step in starting a fresh Operator project is to initialize an empty project structure. First, create an empty project directory with `mkdir nginx-operator` and `cd` into it. Now, initialize a boilerplate project structure with the following:

```
operator-sdk init --domain example.com --repo github.com/
example/nginx-operator
```

> **Note**
>
> This command may take a few moments to complete the first time it is run.

This command sets up a lot of different files and folders that will be filled in with the custom APIs and logic for the Operator we are building. The once-empty project directory should now contain the following files:

```
~/nginx-operator$ ls
total 112K
drwxr-xr-x   12 mdame  staff   384 Dec 22 21:07 .
drwxr-xr-x+ 282 mdame  staff  8.9K Dec 22 21:06 ..
drwx------    8 mdame  staff   256 Dec 22 21:07 config
drwx------    3 mdame  staff    96 Dec 22 21:06 hack
-rw-------    1 mdame  staff   129 Dec 22 21:06 .dockerignore
-rw-------    1 mdame  staff   367 Dec 22 21:06 .gitignore
-rw-------    1 mdame  staff   776 Dec 22 21:06 Dockerfile
-rw-------    1 mdame  staff  8.7K Dec 22 21:07 Makefile
-rw-------    1 mdame  staff   228 Dec 22 21:07 PROJECT
-rw-------    1 mdame  staff   157 Dec 22 21:07 go.mod
-rw-r--r--    1 mdame  staff   76K Dec 22 21:07 go.sum
-rw-------    1 mdame  staff  2.8K Dec 22 21:06 main.go
```

The purposes of these files are as follows:

- `config` – A directory that holds YAML definitions of Operator resources.
- `hack` – A directory that is used by many projects to hold various `hack` scripts. These are scripts that can serve a variety of purposes but are often used to generate or verify changes (often employed as part of a continuous integration process to ensure code is properly generated before merging).

- `.dockerignore` / `.gitignore` – Declarative lists of files to be ignored by Docker builds and Git, respectively.

- `Dockerfile` – Container image build definitions.

- `Makefile` – Operator build definitions.

- `PROJECT` – File used by Kubebuilder to hold project config information (`https://book.kubebuilder.io/reference/project-config.html`).

- `go.mod` / `go.sum` – Dependency management lists for `go mod` (already populated with various Kubernetes dependencies).

- `main.go` – The entry point file for the Operator's main functional code.

With this boilerplate project structure initialized, it is possible to start building Operator logic on top. While this bare project will compile, it doesn't do much besides start an empty controller with `Readyz` and `Healthz` endpoints. To get it to do a little more, first, the Operator must have a defined API.

Defining an API

The Operator's API will be the definition of how it is represented within a Kubernetes cluster. The API is directly translated to a generated CRD, which describes the blueprint for the custom resource object that users will consume to interact with the Operator. Therefore, creating this API is a necessary first step before writing other logic for the Operator. Without this, there will be no way for the Operator's logic code to read values from the custom resource.

Building an Operator API is done by writing a Go struct to represent the object. The basic outline of this struct can be scaffolded by the Operator SDK with the following command:

```
operator-sdk create api --group operator --version v1alpha1
--kind NginxOperator --resource --controller
```

This command does the following:

1. Creates the API types in a new directory called `api/`

2. Defines these types as belonging to the API group `operator.example.com` (since we initialized the project under the domain `example.com`)

3. Creates the initial version of the API named `v1alpha1`

4. Names these types after our Operator, `NginxOperator`

5. Instantiates boilerplate controller code under a new directory called
    `controllers/` (which we will work with more under *Writing a control loop*)

6. Updates `main.go` to add boilerplate code for starting the new controller

For now, we are only concerned with the API types under `api/v1alpha1/`
`nginxoperator_types.go`. There are two other files in this directory
(`groupversion_info.go` and `zz_generated.deepcopy.go`) that do not usually
need to be modified. In fact, the `zz_generated.` prefix is used as a standard to denote
generated files that should never be manually modified. The `groupversion_info.`
`go` file is used to define package variables for this API that instruct clients how to handle
objects from it.

Looking at `nginxoperator_types.go`, there are already some empty structs with
instructions to fill in additional fields. The three most important types in this file are
`NginxOperator`, `NginxOperatorSpec`, and `NginxOperatorStatus`:

```go
// NginxOperatorSpec defines the desired state of NginxOperator
type NginxOperatorSpec struct {
    // INSERT ADDITIONAL SPEC FIELDS - desired state of cluster
    // Important: Run "make" to regenerate code after modifying
this file
    // Foo is an example field of NginxOperator. Edit
nginxoperator_types.go to remove/update
    Foo string `json:"foo,omitempty"`
}
// NginxOperatorStatus defines the observed state of
NginxOperator
type NginxOperatorStatus struct {
    // INSERT ADDITIONAL STATUS FIELD - define observed state of
cluster
    // Important: Run "make" to regenerate code after modifying
this file
}
//+kubebuilder:object:root=true
//+kubebuilder:subresource:status
// NginxOperator is the Schema for the nginxoperators API
type NginxOperator struct {
    metav1.TypeMeta   `json:",inline"`
    metav1.ObjectMeta `json:"metadata,omitempty"`
```

```
Spec    NginxOperatorSpec    `json:"spec,omitempty"`
Status  NginxOperatorStatus  `json:"status,omitempty"`
}
```

As discussed in *Chapter 3, Designing an Operator – CRD, API, and Target Reconciliation*, all Kubernetes API objects should have `Spec` and `Status` fields, and Operators are no different. Therefore, `NginxOperatorSpec` and `NginxOperatorStatus` are those fields, which will be used to accept user input and report on the current state of the Operator, respectively. With `NginxOperator` representing the main object, the relationship between the three is hierarchical.

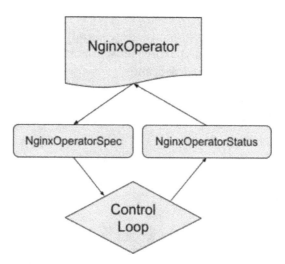

Figure 4.1 – The relationship between NginxOperator fields and logic

Recalling the problem definition from *Chapter 3, Designing an Operator – CRD, API, and Target Reconciliation*, this Operator needs to accept the following inputs:

1. `port`, which will define the port number to expose on the Nginx Pod

2. `replicas`, which defines the number of Pod replicas in order to allow scaling of this deployment through the Operator

3. `forceRedploy`, which is a **no-operation (no-op)** field that simply instructs the Operator to redeploy the `Nginx` Operand

To implement these fields, we need to update the preceding code to modify NginxOperatorSpec with these new fields, as in the following. We are using pointers for the integer fields so that our Operator will be able to distinguish between a zero-value and an unset value, which should fall back to using a defined default:

```
// NginxOperatorSpec defines the desired state of NginxOperator
type NginxOperatorSpec struct {
    // Port is the port number to expose on the Nginx Pod
    Port *int32 `json:"port,omitempty"`

    // Replicas is the number of deployment replicas to scale
    Replicas *int32 `json:"replicas,omitempty"`

    // ForceRedploy is any string, modifying this field
    // instructs  the Operator to redeploy the Operand
    ForceRedploy string `json:"forceRedploy,omitempty"`
}
```

(Note, we also removed the sample Foo field that was generated by the Operator SDK as an example.)

> **Regenerating Code**
>
> Once the Operator types have been modified, it is sometimes necessary to run make generate from the project root. This updates generated files, such as the previously mentioned zz_generated.deepcopy.go. It is good practice to develop the habit of regularly running this command whenever making changes to the API, even if it does not always produce any changes. It is even better practice to add pre-submit continuous integration checks to the Operator's repository to ensure that any incoming code includes these generated changes. Such an automated check can be implemented by running make generate followed by a simple git diff command to assess whether any changes have appeared. If so, the check should fail and instruct the developer to regenerate their code.

For all three new fields, we have also added JSON tags in the form of `json:"...,omitempty"`. The first part of each of these tags defines how the field will appear when represented in JSON or YAML (when interacting with the object through kubectl, for example). omitempty specifies that if this field is empty, it should not show up in JSON output. This is good for hiding optional fields in order to provide a concise output when viewing the objects in the cluster (otherwise, empty fields will appear as nil or with an empty string).

We are going to initially make all three of these fields optional, with default values defined in the Operator. However, they could be designated as required with the removal of omitempty and the addition of more Kubebuilder tags, for example:

```
// Port is the port number to expose on the Nginx Pod
// +kubebuilder:default=8080
// +kubebuilder:validation:Required
Port int `json:"port"`
```

With these settings, any attempt to modify an NginxOperator object without including the port field will result in an error from the API server. In current versions of Kubebuilder, the default assumption is that any field that is not designated as omitempty is required. However, there are ways to switch this default globally (with the // +kubebuilder:validation:Optional marker applied to the top level of an API). Therefore, whenever changing the requirement of a field, it is good practice to explicitly update that field's specific requirement value.

With the API types defined, it is now possible to generate an equivalent CRD manifest, which will be used to create objects matching these types in a Kubernetes cluster.

Adding resource manifests

The relevant resources for an Operator are important to package in a way that can be easily deployed and maintained. This includes the Operator's CRD, but also other resources such as **ClusterRoles** and the matching **ServiceAccount** for those Roles. However, the first step is to generate a CRD from the Go types defined in the previous section with the following:

```
$ make manifests
```

This command generates a CRD that is based on the API we just defined. That CRD is placed under `config/crd/bases/operator.example.com_nginxoperators. yaml`. That CRD looks as follows:

```
apiVersion: apiextensions.k8s.io/v1
kind: CustomResourceDefinition
metadata:
  annotations:
    controller-gen.kubebuilder.io/version: v0.7.0
  creationTimestamp: null
  name: nginxoperators.operator.example.comspec:
  group: operator.example.com
  names:
    kind: NginxOperator
    listKind: NginxOperatorList
    plural: nginxoperators
    singular: nginxoperator
  scope: Namespaced
  versions:
  - name: v1alpha1
    schema:
      openAPIV3Schema:
        description: NginxOperator is the Schema for the
                     nginxoperators API
        properties:
          apiVersion:
            description: 'APIVersion defines the versioned
                         schema of this representation
                         of an object. Servers should convert
                         recognized schemas to the latest
                         internal value, and may reject
                         unrecognized values. More info:
https://git.k8s.io/community/contributors/devel/sig-
architecture/api-conventions.md#resources'
            type: string
          kind:
            description: 'Kind is a string value representing
```

```
                             the REST resource this
                             object represents. Servers may infer
                             this from the endpoint the client
                             submits requests to. Cannot be
                             updated. In CamelCase. More
info: https://git.k8s.io/community/contributors/devel/sig-
architecture/api-conventions.md#types-kinds'
             type: string
          metadata:
             type: object
          spec:
             description: NginxOperatorSpec defines the desired
                          state of NginxOperator
             properties:
               forceRedploy:
                 description: ForceRedploy is any string,
                              modifying this field instructs
                              the Operator to redeploy the
                              Operand
                 type: string
               port:
                 description: Port is the port number to expose
                              on the Nginx Pod
                 type: integer
               replicas:
                 description: Replicas is the number of
                              deployment replicas to scale
                 type: integer
             type: object
          status:
             description: NginxOperatorStatus defines the
                          observed state of NginxOperator
             type: object
          type: object
        served: true
        storage: true
```

```
    subresources:
      status: {}
  status:
    acceptedNames:
      kind: ""
      plural: ""
    conditions: []
    storedVersions: []
```

(In this output, we have added additional formatting to more clearly represent longer strings, such as field descriptions, and highlighted the three fields that were added to the API.)

This CRD is fairly simple due to the basic structure of the Operator, but even that is a testament to the natural complexity of OpenAPI validation schemas. That complexity emphasizes the point that CRD manifests should always be generated and not manually edited.

Customizing Generated Manifests

The default command for generating manifests creates a complex validation schema that should not be edited by hand. However, the underlying command for make manifests is actually calling an additional tool, **controller-gen**. This tool is part of a suite of generators available at https://github.com/kubernetes-sigs/controller-tools. Running controller-gen manually is an acceptable way to generate files and code in non-default ways. For example, the controller-gen schemapatch command will regenerate only the OpenAPI validation schema for a CRD. This can be useful if you wish to manually modify other parts of the CRD, which would be overwritten with a full regeneration, such as additional annotations or labels. The full list of commands can be found by installing controller-gen from the previously mentioned repository and running it with controller-gen -h.

The make manifests command also creates a corresponding **Role-Based Access Control** (**RBAC**) role with can be bound to the Operator's ServiceAccount to give the Operator access to their own custom object:

config/rbac/role.yaml:

```yaml
apiVersion: rbac.authorization.k8s.io/v1
kind: ClusterRole
metadata:
  creationTimestamp: null
  name: manager-role
rules:
- apiGroups:
  - operator.example.com
  resources:
  - nginxoperators
  verbs:
  - create
  - delete
  - get
  - list
  - patch
  - update
  - watch
- apiGroups:
  - operator.example.com
  resources:
  - nginxoperators/finalizers
  verbs:
  - update
- apiGroups:
  - operator.example.com
  resources:
  - nginxoperators/status
  verbs:
  - get
```

```
    - patch
    - update
```

This Role grants full access to create, delete, get, list, patch, update, and watch all `nginxoperator` objects in the cluster. It is generally not a best practice for an Operator to manage the life cycle of their own custom resource object (for example, creation of the `config` object is best left to a manual action by the user), so some of the verbs, such as `create` and `delete`, are not strictly necessary here. However, we will leave them for now.

Additional manifests and BinData

The remaining resource manifests that will need to be created include the Operator's `ClusterRole` and the Nginx Deployment definition. The `ClusterRole` can be conveniently generated with Kubebuilder tags in the code, which will be done later in the *Writing a control loop* section. Before that, the Deployment should be defined so that the control loop has access to it.

One way to define in-memory resources, such as the Deployment, is by creating them in the code. Many projects take such an approach, including the official example projects available at `https://github.com/operator-framework/operator-sdk`. For the purpose of this Nginx Deployment, the approach would involve creating a function similar to the following:

```
func nginxDeployment() *appsv1.Deployment {
    dep := &appsv1.Deployment{
        ObjectMeta: metav1.ObjectMeta{
            Name:      "nginxDeployment",
            Namespace: "nginxDeploymentNS",
        },
        Spec: appsv1.DeploymentSpec{
            Replicas: &pointer.Int32Ptr(1),
            Selector: &metav1.LabelSelector{
                MatchLabels: map[string]
string{"app":"nginx"},
            },
            Template: corev1.PodTemplateSpec{
                ObjectMeta: metav1.ObjectMeta{
                    Labels: map[string]
string{"app":"nginx"},,               },
```

```
            Spec: corev1.PodSpec{
        Containers: []corev1.Container{{
                                    Image:    "nginx:latest",
                                    Name:     "nginx",
                                    Command:  []string{"nginx"
      Ports: []corev1.ContainerPort{{
                                      ContainerPort: 8080,
                                        Name:          "nginx",
    }},                                 }},
                        },
                },              },
            }
        return dep
    }
```

The preceding function returns a static Deployment Go struct, with default values prepopulated, such as the Deployment name and exposed port. This object could then be modified based on the specifications set in the Operator CRD before using a Kubernetes API client to update the Deployment in the cluster (for example, to update the number of replicas).

As discussed in *Chapter 3, Designing an Operator – CRD, API, and Target Reconciliation*, this approach is easy to code because the resource struct is immediately available as a Go type. However, in terms of maintainability and usability, there are better options. This is where tools such as **go-bindata** and **go:embed** are helpful.

Using go-bindata and go:embed to access resources

The go-bindata project is available on GitHub at `https://github.com/go-bindata/go-bindata`. It works by converting arbitrary files into Go code, which is then compiled into the main program and available in memory. The benefit of using go-bindata is that project resources can be more concisely managed and maintained in a more readable format such as YAML, which provides familiarity with native Kubernetes resource creation. Since Go 1.16, the language has included its own compiler directive, `go:embed`, to essentially perform the same function; however, we will provide an example of both approaches for the benefit of users who have not yet updated to Go 1.16 or who wish to avoid reliance on compiler-specific directives in their development and production environments.

The first step for either approach is to create the resource manifests in a directory, such as `assets/nginx_deployment.yaml`:

```yaml
apiVersion: apps/v1
kind: Deployment
metadata:
  name: "nginx-deployment"
  namespace: "nginx-operator-ns"
  labels:
    app: "nginx"
spec:
  replicas: 1
  selector:
    matchLabels:
      app: "nginx"
  template:
    metadata:
      labels:
        app: "nginx"
    spec:
      containers:
        - name: "nginx"
          image: "nginx:latest"
          command: ["nginx"]
```

This structure is already much easier to work with than the native Go types by saving us from having to define each embedded type (such as `map[string]string` for the labels and Pod command). It can also be easily parsed by continuous integration checks to ensure it maintains a valid structure.

The next two subsections will demonstrate the basic concepts of implementing either `go-bindata` or `go:embed`. These examples will do so by showing how you could add the foundational concepts for each approach. However, we will ultimately refactor most of the code in the section titled *Simplifying resource embedding*; therefore, you may choose not to write any of this code yourself until reaching the refactoring subsection.

Go 1.15 and older – go-bindata

For older versions of Go, you must install the `go-bindata` package from GitHub to generate your files:

```
$ go get -u github.com/go-bindata/go-bindata/...
```

The generated code that contains the manifests can then be created with the following command:

```
$ go-bindata -o assets/assets.go assets/...
```

This command will create an `assets.go` file under the `assets/` directory that contains generated functions and in-memory representations of the files in the `assets/` directory. Note that it can be easier to keep the assets themselves in a different directory than the generated code, as re-running the `go-bindata` command will now include a representation of the `assets.go` file itself unless it is excluded, like so:

```
$ go-bindata -o assets/assets.go -ignore=\assets\/assets\.go
assets/…
```

This file will need to be re-generated whenever modifications are made to the underlying assets. Doing so will ensure the changes are made available in the compiled assets package that includes the accessibility functions for the files.

Once the assets have been generated, they can be accessed in code by importing the new `assets` package and using the `Asset()` function like so:

```
import "github.com/sample/nginx-operator/assets"

...

asset, err := assets.Asset("nginx_deployment.yaml")
if err != nil {
   // process object not found error
}
```

For newer versions of Go (1.16 and above), it is even simpler to compile resource manifests into assets.

Go 1.16 and newer – go:embed

The `go:embed` marker was included as a compiler directive in Go 1.16 to provide native resource embedding without the need for external tools such as go-bindata. To start with this approach, create the resource manifest files similarly to the go-bindata setup under a new directory called `assets/`.

Next, the `main.go` file for the Operator project needs to be modified to import the embed package and declare the asset manifests as a variable like so (all of the following code in this section shows only the changes that you need to make):

```
package main
import (

   ...

   "embed"

   ...

)


//go:embed assets/nginx_deployment.yaml
var deployment embed.FS
```

Note the `//go:embed` comment, which tells the compiler to store the contents of `assets/nginx_deployment.yaml` as filesystem data in the `deployment` variable.

The data can then be read and converted to a Deployment Go struct by utilizing the Kubernetes API scheme declarations like so:

```
import (
   appsv1 "k8s.io/api/apps/v1"
   "k8s.io/apimachinery/pkg/runtime"
   "k8s.io/apimachinery/pkg/runtime/serializer"
)
...
var (
       appsScheme = runtime.NewScheme()
       appsCodecs = serializer.NewCodecFactory(appsScheme)
)


...
func main() {

if err := appsv1.AddToScheme(appsScheme); err != nil {
   panic(err)
}
```

```
deploymentBytes, err := deployment.ReadFile("assets/nginx_
deployment.yaml")
if err != nil {
  panic(err)
}


deploymentObject, err := runtime.Decode(appsCodecs.
UniversalDecoder(appsv1.SchemeGroupVersion), deploymentBytes)
if err != nil {
  panic(err)
}
dep := deploymentObject.(*appsv1.Deployment)
...
}
```

This code does a few things:

1. It imports the relevant Kubernetes API packages that define the schema for
    Deployment API objects:

    ```
    import (
      appsv1 "k8s.io/api/apps/v1"
      "k8s.io/apimachinery/pkg/runtime"
      "k8s.io/apimachinery/pkg/runtime/serializer"
    )
    ```

2. It initializes a Scheme and a set of codecs that can be used by the API's
    UniversalDecoder in order to know how to convert the [] byte data representation
    of the file to a Go struct:

    ```
    var (
        appsScheme = runtime.NewScheme()
        appsCodecs = serializer.NewCodecFactory(appsScheme)
    )
    ```

3. It uses the deployment variable we declared earlier (as part of setting up
    the embed directive) to read the Deployment file under assets/nginx_
    deployment.yaml (highlighted):

    ```
    deploymentBytes, err := deployment.ReadFile("assets/
    nginx_deployment.yaml")
    ```

```
if err != nil {
  panic(err)
}
```

4. It decodes the `[]byte` data returned from `deployment.ReadFile()` into an object that can be cast to the Go type for Deployments:

```
deploymentObject, err := runtime.Decode(appsCodecs.
UniversalDecoder(appsv1.SchemeGroupVersion),
deploymentBytes)
if err != nil {
  panic(err)
}
```

5. It casts the object data to an in-memory representation of `*appsv1.` `Deployment`:

```
dep := deploymentObject.(*appsv1.Deployment)
```

From this point, it will be necessary to find a way to pass the Deployment object to our Nginx Operator Controller. This can be done by modifying the `NginxOperatorReconciler` type to have a field that holds a type of `*appsv1.` `Deployment`. However, this is not convenient for all of the different types of resources the Operator will be managing. To simplify this and better organize the project's structure, we can move the resource embedding code to its own package.

Simplifying resource embedding

The preceding examples showed the essential steps of embedding a YAML file into Go code. However, for our sample Nginx Operator, this can be better organized into its own package. To do this, we will keep the existing `assets/` directory (to use as an importable Go module path that holds helper functions for loading and processing the files) and place a new `manifests/` directory underneath it (which will hold the actual manifest files). This new file structure will look as follows:

```
./nginx-operator/
| - assets/
| - - assets.go
| - - manifests/
| - - - nginx_deployment.yaml
```

The assets.go file will include the API schema initialization and wrapped object casting functionality from the preceding examples, like this:

```go
package assets
import (
    "embed"
    appsv1 "k8s.io/api/apps/v1"
    "k8s.io/apimachinery/pkg/runtime"
    "k8s.io/apimachinery/pkg/runtime/serializer"
)
var (
    //go:embed manifests/*
    manifests embed.FS

    appsScheme = runtime.NewScheme()
    appsCodecs = serializer.NewCodecFactory(appsScheme)
)
func init() {
    if err := appsv1.AddToScheme(appsScheme); err != nil {
        panic(err)
    }
}
func GetDeploymentFromFile(name string) *appsv1.Deployment {
    deploymentBytes, err := manifests.ReadFile(name)
    if err != nil {
        panic(err)
    }
    deploymentObject, err := runtime.Decode(
        appsCodecs.UniversalDecoder(appsv1.SchemeGroupVersion),
        deploymentBytes,
    )
    if err != nil {
        panic(err)
    }
    return deploymentObject.(*appsv1.Deployment)
}
```

This new file makes a couple of changes from the implementation shared before:

1. Now, the entire `manifests/` directory is embedded as a filesystem variable. This will make it easier to add functions to read other resources within the directory without having to declare new variables in this package for each one.

2. The main logic has been wrapped in a new function, `GetDeploymentFromFile()`. This function can be called and used by our control loop like this:

```
import "github.com/sample/nginx-operator/assets"

...

nginxDeployment := assets.
GetDeploymentFromFile("manifests/nginx_deployment.yaml")
```

We can add other manifests to this directory so that the Operator can manage them (for example, additional Operand dependencies). But for now, we have enough to begin working on a control loop.

Writing a control loop

With a strategy for the in-memory representation of resource manifests in place, it is now much easier to begin writing the Operator's control loop. As described in previous chapters, this control loop comprises a core state reconciliation function call that is triggered by certain relevant cluster events. This function does not run continuously on a loop, but rather the main thread of the Operator will be continuously observing the cluster for those events to kick off a call to the state reconciliation function.

The empty `Reconcile()` function has been scaffolded already by the Operator SDK in `controllers/nginxoperator_controller.go`:

```
func (r *NginxOperatorReconciler) Reconcile(ctx context.
Context, req ctrl.Request) (ctrl.Result, error) {
    _ = log.FromContext(ctx)
    // your logic here
    return ctrl.Result{}, nil

}
```

Right now, this function does nothing but return an empty `ctrl.Result` and an empty `error`, which evaluates to a successful run and instructs the rest of the framework that there is no need to re-try this reconciliation run. If this function returns either a non-nil `error` or non-empty `ctrl.Result` struct, then the controller will instead re-queue this reconciliation attempt to be tried again. These cases will come up as we populate the controller's logic as indicated by the comment.

Because the Operator SDK instantiates this controller with a Kubernetes client already accessible, we can use functions such as `Get()` to retrieve cluster resources. The first thing to do is to attempt to access the existing Nginx Operator resource object. If one is not found, we should log a message indicating so and terminate the reconciliation attempt. If there are any other errors retrieving the object, we will instead return an error, so this attempt gets re-queued and tried again. This approach can account for other failures, such as network issues or temporarily dropped connections to the API server. With these changes, the new `Reconcile()` function looks like the following:

controllers/nginxoperator_controller.go:

```go
func (r *NginxOperatorReconciler) Reconcile(ctx context.
Context, req ctrl.Request) (ctrl.Result, error) {
  logger := log.FromContext(ctx)
  operatorCR := &operatorv1alpha1.NginxOperator{}
  err := r.Get(ctx, req.NamespacedName, operatorCR)
  if err != nil && errors.IsNotFound(err) {
    logger.Info("Operator resource object not found.")
    return ctrl.Result{}, nil
  } else if err != nil {
    logger.Error(err, "Error getting operator resource object")
    return ctrl.Result{}, err
  }
  return ctrl.Result{}, nil
}
```

This error-handling pattern is common among Kubernetes projects, including examples from the official Operator SDK documentation. Rather than immediately returning the error, this code essentially ignores the case where the Operator object is simply not found. The `errors.IsNotFound()` check comes from the `k8s.io/apimachinery/pkg/api/errors` package, which provides several helper functions for a standard approach to handling specific Kubernetes errors. Using this pattern helps to minimize logging noise for the user and to ignore events where the Operator resource was deleted (which will still trigger a reconciliation attempt). In the event that the Operator object is not found, developers can go a step further and use that as a signal to take other steps (such as deleting other resources that depend on the Operator deployment).

Also, note that this code is using `req.NamespacedName` to get the `Name` and `Namespace` of the Operator config object. This follows one of the best practices laid out in the Operator Framework documentation (`https://sdk.operatorframework.io/docs/best-practices/best-practices/`):

Operators shouldn't make any assumptions about the namespace they are deployed in, and they should not use hardcoded names of resources that they expect to already exist.

In this case, the `req` parameter includes the name of the object event that triggered the reconciliation attempt. Using a consistent name across resources allows us to re-use the `req.NamespacedName` field in every call to `Reconcile()`, regardless of the object that triggered the reconciliation. In other words, if the Deployment has the same Name and Namespace as the Operator custom resource object, we can consistently eliminate the use of hardcoded assumptions for resource names.

With an Operator resource object successfully found, the controller can retrieve the values for each setting in the Operator's `spec` in order to update them, if necessary. Similar to what we just did for the Operator resource, however, we must first check whether the Deployment exists. For this, we will follow a similar pattern utilizing `errors.IsNotFound()` to check for the non-existence of an Nginx Deployment. However, in this case, the function will not simply return if no Deployment is found, but instead the controller will create one from the embedded Deployment YAML file:

controllers/nginxoperator_controller.go:

```go
deployment := &appsv1.Deployment{}
err = r.Get(ctx, req.NamespacedName, deployment)
if err != nil && errors.IsNotFound(err) {
  deployment.Namespace = req.Namespace
  deployment.Name = req.Name
```

```
   deploymentManifest := assets.
GetDeploymentFromFile("manifests/nginx_deployment.yaml")
   deploymentManifest.Spec.Replicas = operatorCR.Spec.Replicas
deploymentManifest.Spec.Template.Spec.Containers[0].Ports[0].
ContainerPort = *operatorCR.Spec.Port
   err = r.Create(ctx, deploymentManifest)
   if err != nil {
      logger.Error(err, "Error creating Nginx deployment.")
      return ctrl.Result{}, err
   }
return ctrl.Result{}, nil
} else if err != nil {
   logger.Error(err, "Error getting existing Nginx deployment.")
   return ctrl.Result{}, err
}
```

In this code, we are loading the embedded default manifest using the `assets.GetDeploymentFromFile()` function created earlier. We are also modifying that manifest declaration to include the values from the current Operator resource object.

An alternative to this approach would be to create the Deployment with the default values, and then have the function immediately return with `ctrl.Result{Requeue: true}`. This would trigger another reconciliation attempt, where the Deployment should be found and then updated with the Operator resource settings. The trade - off here is the immediate creation of a new object without the need for another reconciliation cycle, in exchange for less atomic operation and some code duplication (because we will still need the following code to apply the Operator resource settings in the case where an existing Deployment was found). To eliminate that duplicate code, we can modify the preceding section like this:

controllers/nginxoperator_controller.go:

```
deployment := &appsv1.Deployment{}
create := false
err = r.Get(ctx, req.NamespacedName, deployment)
   if err != nil && errors.IsNotFound(err) {
      create = true
      deployment = assets.GetDeploymentFromFile("manifests/nginx_
deployment.yaml")
```

```
} else if err != nil {
  logger.Error(err, "Error getting existing Nginx deployment.")
  return ctrl.Result{}, err
}
if operatorCR.Spec.Replicas != nil {
  deployment.Spec.Replicas = operatorCR.Spec.Replicas
}

if operatorCR.Spec.Port != nil {
deployment.Spec.Template.Spec.Containers[0].Ports[0].
ContainerPort = *operatorCR.Spec.Port
}
```

Since the `Replicas` and `Port` fields in the Operator custom resource are optional (and pointers), we should use `nil` checks to see whether any value has been set. Otherwise, the Deployment will default to the values defined in its `manifest` file.

Now, we are always making sure that the Deployment object is being modified to include the Operator settings, whether it is the existing Deployment or a new one. Then, the decision to call `Create()` or `Update()` will be made later based on the value of the `create` Boolean:

controllers/nginxoperator_controller.go:

```
if create {
  err = r.Create(ctx, deployment)
} else {
  err = r.Update(ctx, deployment)
}
return ctrl.Result{}, err
```

If either call results in an error, it will be returned and logged by the controller thanks to the scaffolded framework code. If the call to create or update the Deployment is successful, then `err` will be `nil` and the reconciliation call completes successfully. Our full `Reconcile()` function now looks like this:

controllers/nginxoperator_controller.go:

```
func (r *NginxOperatorReconciler) Reconcile(ctx context.
Context, req ctrl.Request) (ctrl.Result, error) {
```

```
    logger := log.FromContext(ctx)


    operatorCR := &operatorv1alpha1.NginxOperator{}
    err := r.Get(ctx, req.NamespacedName, operatorCR)
    if err != nil && errors.IsNotFound(err) {
        logger.Info("Operator resource object not found.")
        return ctrl.Result{}, nil
    } else if err != nil {
        logger.Error(err, "Error getting operator resource
object")
        return ctrl.Result{}, err
    }


    deployment := &appsv1.Deployment{}
    create := false
    err = r.Get(ctx, req.NamespacedName, deployment)
    if err != nil && errors.IsNotFound(err) {
        create = true
        deployment = assets.GetDeploymentFromFile("assets/nginx_
deployment.yaml")
    } else if err != nil {
        logger.Error(err, "Error getting existing Nginx
deployment.")
        return ctrl.Result{}, err
    }


    deployment.Namespace = req.Namespace
    deployment.Name = req.Name

    if operatorCR.Spec.Replicas != nil {
        deployment.Spec.Replicas = operatorCR.Spec.Replicas
    }


    if operatorCR.Spec.Port != nil { deployment.Spec.Template.
Spec.Containers[0].Ports[0].ContainerPort = *operatorCR.Spec.
Port
    }
```

```
ctrl.SetControllerReference(operatorCR, deployment, r.Scheme)

    if create {
        err = r.Create(ctx, deployment)
    } else {
        err = r.Update(ctx, deployment)
    }
    return ctrl.Result{}, err
}
```

In addition, we added a call to `ctrl.SetControllerReference()` to indicate that the Nginx Operator resource object should be listed as the `OwnerReference` (an API field denoting which object "owns" the specified object) of the Nginx Deployment, which helps with garbage collection.

Finally, we need to ensure that the Operator actually has the cluster permissions necessary to get, create, and update Deployments. To do that, we need to update the RBAC role for the Operator. This can be done automatically using Kubebuilder markers on the `Reconcile()` function, which helps keep permissions organized and their necessary usage clearly identified. There are already Kubebuilder markers that were generated for accessing the Operator custom resource, but now we can add additional ones for Deployments:

---

controllers/nginxoperator_controller.go:

```
//+kubebuilder:rbac:groups=operator.example.com,
resources=nginxoperators,verbs=get;list;watch;create;
update;patch;delete
//+kubebuilder:rbac:groups=operator.example.com,
resources=nginxoperators/status,verbs=get;update;patch
//+kubebuilder:rbac:groups=operator.example.com,
resources=nginxoperators/finalizers,verbs=update
//+kubebuilder:rbac:groups=apps,resources=deployments,
verbs=get;list;watch;create;update;patch;delete
```

Now, running `make manifests` should produce this new section in the Operator's ClusterRole (`config/rbac/role.yaml`):

```
rules:
- apiGroups:
  - apps
```

```
resources:
- deployments
verbs:
- create
- delete
- get
- list
- patch
- update
- watch
```

At this point, we now have a basic control loop reconciling the specified Operator settings with the current state of the cluster. But what events will trigger this loop to run? This is set up in `SetupWithManager()`:

controllers/nginxoperator_controller.go:

```go
// SetupWithManager sets up the controller with the Manager.
func (r *NginxOperatorReconciler) SetupWithManager(mgr ctrl.
Manager) error {
    return ctrl.NewControllerManagedBy(mgr).
        For(&operatorv1alpha1.NginxOperator{}).
        Complete(r)
}
```

This code was generated to observe the cluster for changes to the `NginxOperator` objects, but we need it to also observe changes to Deployment objects (since the Operator is managing a Deployment). This can be done by modifying the function like this:

```go
// SetupWithManager sets up the controller with the Manager.
func (r *NginxOperatorReconciler) SetupWithManager(mgr ctrl.
Manager) error {
    return ctrl.NewControllerManagedBy(mgr).
        For(&operatorv1alpha1.NginxOperator{}).
        Owns(&appsv1.Deployment{}).
        Complete(r)
}
```

With the added call to `Owns(&appsv1.Deployment{})`, the controller manager now knows to also trigger calls to `Reconcile()` for changes to Deployment objects in the cluster. The controller will now evaluate any changes to a Deployment object as a relevant event for the `NginxOperator` object since we have listed it as the owner of the Deployment. Multiple types of objects can be chained into this watch list with subsequent calls to `Owns()`.

We now have an Operator that does something. When built and deployed, this controller will watch for changes to any `NginxOperator` custom resource in the cluster and react to them. This means that the first time a user creates the Operator custom resource object, the Operator will see that a configuration object now exists, and create a Deployment based on the values present. It will also observe for changes to Deployments. For example, if the Nginx Deployment is accidentally deleted, the Operator will respond by creating a new one with the existing settings from its custom resource.

Troubleshooting

The steps outlined in this chapter involve using several different tools and libraries with varying dependency requirements. Understandably, this can lead to errors, especially in different development environments. While the authors of the software involved have taken steps to produce informative and helpful error messages when necessary, it is not always possible to provide a clear resolution with automated responses. Such is the unfortunate nature of rapidly evolving software development.

Fortunately, however, the benefits of open source software provide many resources and volunteers to help support and debug issues, should they arise. This section will highlight those resources as a guide for resolving technical issues. All of these tools offer documentation and how to guides, but many of them also have community resources where users can ask for clarification and assistance from maintainers and other users.

General Kubernetes resources

The Operator SDK is foundationally built upon several Kubernetes libraries; therefore, it is very helpful to understand some of the Kubernetes packages that are used to build Operators with the framework. By doing so, it can sometimes be easier to find the root cause of an issue.

The Kubernetes reference documents are located at `https://kubernetes.io/docs/home/`. However, this home section is mostly oriented toward usage documentation. For support regarding the Kubernetes API (including API clients, standards, and object references), the API reference section is much more relevant to the topics covered in this chapter. That is located at `https://kubernetes.io/docs/reference/`.

The entire Kubernetes source code is available on GitHub under various organizations, such as `https://github.com/kubernetes` (most of the project code) and `https://github.com/kubernetes-sigs` (subprojects, such as Kubebuilder). For example, the Go client library, which is used by the Operator SDK framework under the hood to provide resource functions such as `r.Get()`, is hosted at `https://github.com/kubernetes/client-go/`.

Familiarity with the GitHub repositories that host the different code dependencies that the Operator Framework is built on provides an excellent resource for communication with the maintainers of these projects. Searching the *Issues* on GitHub can very often provide immediate relief to an issue (or at least provide insight into the current status of ongoing problems).

For faster responses and a wider audience, the Kubernetes community is very active on the Slack messaging platform. The official Kubernetes Slack server is open to anyone, at `slack.k8s.io`. Helpful channels for developers working with general Kubernetes issues include the following:

- **#kubernetes-novice** – This channel is for new Kubernetes users and developers.

- **#kubernetes-contributors** – This channel is more dedicated to the development of Kubernetes itself, but there are still useful topics covered around relevant topics, such as API clients.

- **#kubernetes-users** – Similar to `#kubernetes-novice` with a focus on usage rather than development, but for more specific questions.

- **#sig-api-machinery** – The different areas and subprojects of Kubernetes are owned by their respective **Special Interest Groups** (**SIGs**). `SIG-APIMachinery` is responsible for the ownership of the Kubernetes Go client, which we used by extension in this chapter. Here you will find the most knowledgeable contributors in regard to API topics.

For the topics in this chapter, these resources are relevant for issues related to the Kubernetes API, including the generated client tools created with commands such as `make generate`.

Operator SDK resources

The Operator SDK also provides a wealth of documentation, including example Operator development tutorials. In this chapter, developing an Operator in Go meant following similar steps to those outlined in the Operator SDK Go documentation, located at `https://sdk.operatorframework.io/docs/building-operators/golang/`.

Similar to other Kubernetes projects, Operator SDK is also available on GitHub at `https://github.com/operator-framework/operator-sdk/`. This is a great resource for examples, issue tracking, and staying notified of updates and ongoing work with the project.

There are several Operator-specific channels on `slack.k8s.io`, including **#operator-sdk-dev** (which is meant for discussion related to the Operator SDK) and **#kubernetes-operators**, which is for general discussion purposes regarding Operators.

These resources are all helpful for problems related to the `operator-sdk` binary, or the patterns provided by the SDK's code libraries and patterns.

Kubebuilder resources

Kubebuilder is the tool that is used by Operator SDK to generate manifests and some controller code. This includes commands that were run in this chapter, such as `make manifests`, so for most issues related to CRDs or generating them from code markers (for example, `//+kubebuilder...`), this is a good starting point for assistance.

An excellent reference for Kubebuilder is the Kubebuilder Book, available at `https://book.kubebuilder.io/`. This is the essential documentation reference for Kubebuilder and includes details on all of the available comment markers for generating code. Its code base is also available on GitHub at `https://github.com/kubernetes-sigs/kubebuilder`, and some of its sub-tools (such as controller-gen) are available at `https://github.com/kubernetes-sigs/controller-tools`.

Finally, there is the **#kubebuilder** channel on the Kubernetes Slack server for interactive discussion and help with this tool.

Summary

This chapter followed the design we outlined in *Chapter 3, Designing an Operator – CRD, API, and Target Reconciliation*, to produce functional code that achieves the minimum requirements for a *Level I* Operator (Basic Install). With the support of the Operator Lifecycle Manager (which will be demonstrated in later chapters) and good subsequent API design, this Operator will also support upgrades for itself and its Operand, which qualifies it for *Level II*.

The steps for creating a Go-based Operator, as recommended by the Operator SDK documentation approach, build upon each other to achieve base functionality. In this chapter, that pattern meant first designing the Operator's API types, which are then generated into a CRD using tools such as Kubebuilder. At this time, it is good to begin thinking about other resource manifests, such as the Operand Deployment, and how those will be represented in memory. This guide took the approach of embedding these additional resources directly into the Go binary using built-in Go compiler directives that allow the language to do this natively.

Finally, the core controller code was filled in. This is what makes the Operator a controller, and this control loop is used to reconcile the desired state of the cluster with the actual state of the cluster based on user input through the Operator's CRD. With some additional tweaks to the event triggers and added RBAC permissions, this code begins to observe Deployments, which is necessary to manage the Operand.

In the next chapter, we will build on this basic functionality to add more advanced code. This will bring our Operator beyond Level II, as we add things such as metrics and leader election to create a more sophisticated controller capable of deeper insights and error handling.

5
Developing an Operator – Advanced Functionality

While a cluster with Operators that are capable of basic installation and upgrade functionality is a considerable improvement over non-Operator-based Kubernetes clusters, there is still more that can be done to improve cluster administration and user experience. Advanced features can help users to achieve more sophisticated automation, guide failure recovery, and inform data-driven deployment decisions with features such as metrics and status updates.

These are some of the fundamental features for higher-level Operators within the **Capability Model** (as described in *Chapter 1, Introduction to the Operator Framework*). As such, this chapter will first explain the cost and benefits of implementing advanced functionality (in relation to the effort necessary to do so) before demonstrating ways to add these features in the following sections:

- Understanding the need for advanced functionality
- Reporting status conditions
- Implementing metrics reporting

- Implementing leader election
- Adding health checks

Conveniently, the code required to implement these features does not require any significant refactoring of the existing Operator code. In fact, the hierarchical nature of the Capability Model and the development patterns provided by the Operator SDK encourage this iterative construction. It is, therefore, the goal of this chapter to build upon the basic Operator code from *Chapter 4*, *Developing an Operator with the Operator SDK*, to create a more complex Operator capable of providing the features we just listed.

Technical requirements

The examples shown throughout this chapter build upon the project code that was started in *Chapter 4*, *Developing an Operator with the Operator SDK*. Therefore, it is recommended to start with that chapter (and the prerequisites for it), which covers project initialization and basic Operator functionality. This is not required, however, and the sections in this chapter can be generally applied to any Operator SDK project. That is, any project initialized by the Operator SDK will work with the following steps and you do not need to specifically implement all of the code from previous chapters.

With that in mind, the requirements for this chapter are as follows:

- Any existing Operator SDK project
- Go 1.16+

The Code in Action video for this chapter can be viewed at: `https://bit.ly/3zbsvD0`

Understanding the need for advanced functionality

With a basic, functional Operator already built and ready for deployment, you may be asking, What else do I really need? Indeed, now that your operand is installable and its health is managed by your Operator, there may be nothing more to do. This is a perfectly acceptable level of functionality for an Operator to have. In fact, it may be preferable to start with a simple Operator and iterate as your development resources allow (recall discussing this in *Chapter 3*, *Designing an Operator – CRD, API, and Target Reconciliation*).

The point is that there is no shame during the development of your Operator in stopping here. The Capability Model defines lower-level Operators for a reason (in other words, if it was unacceptable to have an Operator that can only install an operand, then why would Level I be defined at all?).

However, the Capability Model does define higher-level Operators for a reason too. It is not difficult to imagine, for example, that during the course of a user's interaction with your Operator, they may wish to see more detailed insights into how it performs within a production environment. This is a good use case for adding custom metrics to an Operator. Or, there may be a common failure state that is difficult to debug (wherein status conditions would help expose more information about the failure in an efficient way).

The following sections are just a few of the most common additional features that help elevate an Operator to a higher level of functionality. Some of these are also covered in additional detail by the Operator SDK documentation on *Advanced Topics* (`https://sdk.operatorframework.io/docs/building-operators/golang/advanced-topics/`). It is, of course, not feasible to list every possible feature that you could add to your Operator. But hopefully, these examples serve as a good starting point for your own development.

Reporting status conditions

Status conditions were discussed in *Chapter 3, Designing an Operator – CRD, API, and Target Reconciliation*, as a way to efficiently communicate human-readable information about the Operator's health to administrators. By presenting directly in an Operator's **CustomResourceDefinition** (**CRD**), the information they provide is more easily highlighted and viewable in a more centralized starting point for debugging issues. In this way, they provide an advantage over error logs, which can contain lots of unrelated information and often lack direct context, making them hard to trace to a root cause.

Implementing conditions in an Operator is made easier by the **Kubernetes** API's standardization of the concept. The standard `Condition` type was implemented in *KEP-1623* (`https://github.com/kubernetes/enhancements/tree/master/keps/sig-api-machinery/1623-standardize-conditions`) around Kubernetes 1.19. That type is now part of the Kubernetes API in the `k8s.io/apimachinery/pkg/api/meta` module. This allows developers to work with a consistent understanding of how conditions should be reported in Kubernetes with all of the compatibility assurances of Kubernetes API support.

The Operator Framework has also implemented conditions based on the Kubernetes type, both in the Operator SDK and the **Operator Lifecycle Manager** (**OLM**). Accordingly, conditions can be set either on the Operator's custom resource or on an additional `OperatorCondition` resource that the OLM creates. This section will cover both approaches.

Operator CRD conditions

As part of the Kubernetes API conventions covered in *Chapter 3, Designing an Operator – CRD, API, and Target Reconciliation*, objects (including custom resources) should include both a spec and status field. In the case of Operators, we are using spec as an input for configuring the Operator's parameters already. However, we have not yet modified the status field. We will now change that by adding a list of conditions as a new field in api/v1alpha1/nginxoperator_types.go:

```
// NginxOperatorStatus defines the observed state of
NginxOperator
type NginxOperatorStatus struct {
    // Conditions is the list of status condition updates
    Conditions []metav1.Condition `json:"conditions"`
}
```

Then, we will run make generate (to update the generated client code) and make manifests (to update the Operator's CRD with the new field), or simply make (which runs all generators, though we don't need some of them right now). The new field is now reflected in the CRD:

```
properties:
  conditions:
    description: Conditions is the list of the most recent
status condition
      updates
    items:
      description: "Condition contains details for one aspect
                   of the current state of this API Resource.
                   --- This struct is intended for direct use
                   as an array at the field path
                   .status.conditions."
      properties:
        lastTransitionTime:
          description: lastTransitionTime is the last time the
                       condition transitioned from one status
                       to another. This should be when the
                       underlying condition changed.  If that
                       is not known, then using the time when
```

```
                          the API field changed is acceptable.
            format: date-time
            type: string
        message:
            description: message is a human readable message
                         indicating details about the
                         transition. This may be an empty
                         string.
            maxLength: 32768
            type: string
...

        status:
            description: status of the condition, one of True,
                         False, Unknown.
            enum:
            - "True"
            - "False"
            - Unknown
            type: string
        type:
            description: type of condition in CamelCase or in
                         foo.example.com/CamelCase. --- Many
                         .condition.type values are consistent
                         across resources like Available, but
                         because arbitrary conditions can be
                         useful (see .node.status.conditions),
                         the ability to deconflict is important.
                         The regex it matches is
                         (dns1123SubdomainFmt/)?(qualifiedNameFmt)
            maxLength: 316
            pattern: ^(([a-z0-9]([-a-z0-9]*[a-z0-9])?(\.[a-z0-9]
([-a-z0-9]*[a-z0-9])?)*/)?(([A-Za-z0-9][-A-Za-z0-9_.]*)?[A-
Za-z0-9])$
            type: string
        required:
        - lastTransitionTime
```

```
        - message
        - reason
        - status
        - type
      type: object
    type: array
  required:
  - conditions
  type: object
```

Note that this also imports all of the embedded validation requirements for the `Condition` type from the Kubernetes API.

Now that the Operator's CRD has a field to report the latest status conditions, the code can be updated to implement them. For this, we can use the `SetStatusCondition()` helper function, which is available in the `k8s.io/apimachinery/pkg/api/meta` module. For this example, we will start with a single condition called `OperatorDegraded`, which will default to `False` to indicate that the Operator is successfully reconciling changes in the cluster. If the Operator does encounter an error, however, we will update this condition to `True` with a message indicating the error. This will involve some refactoring of the `Reconcile()` function in `controllers/nginxoperator_controller.go` to match the following:

```
func (r *NginxOperatorReconciler) Reconcile(ctx context.
Context, req ctrl.Request) (ctrl.Result, error) {
    logger := log.FromContext(ctx)


    operatorCR := &operatorv1alpha1.NginxOperator{}
    err := r.Get(ctx, req.NamespacedName, operatorCR)
    if err != nil && errors.IsNotFound(err) {
      logger.Info("Operator resource object not found.")
      return ctrl.Result{}, nil
    } else if err != nil {
      logger.Error(err, "Error getting operator resource
object")
        meta.SetStatusCondition(&operatorCR.Status.Conditions,
metav1.Condition{
            Type:                "OperatorDegraded",
            Status:              metav1.ConditionTrue,
```

```
        Reason:                "OperatorResourceNotAvailable",
        LastTransitionTime: metav1.NewTime(time.Now()),
        Message:               fmt.Sprintf("unable to get
operator custom resource: %s", err.Error()),
    })
    return ctrl.Result{}, utilerrors.NewAggregate([]
error{err, r.Status().Update(ctx, operatorCR)})
  }
```

The preceding code will now attempt to report a degraded status condition if the controller is initially unable to access the Operator's CRD. The code continues as follows:

```
    deployment := &appsv1.Deployment{}
    create := false
    err = r.Get(ctx, req.NamespacedName, deployment)
    if err != nil && errors.IsNotFound(err) {
        create = true
        deployment = assets.GetDeploymentFromFile("assets/nginx_
deployment.yaml")
    } else if err != nil {
        logger.Error(err, "Error getting existing Nginx
deployment.")
        meta.SetStatusCondition(&operatorCR.Status.Conditions,
metav1.Condition{
            Type:               "OperatorDegraded",
            Status:             metav1.ConditionTrue,
            Reason:             "OperandDeploymentNotAvailable",
            LastTransitionTime: metav1.NewTime(time.Now()),
            Message:            fmt.Sprintf("unable to get operand
deployment: %s", err.Error()),
        })
        return ctrl.Result{}, utilerrors.NewAggregate([]
error{err, r.Status().Update(ctx, operatorCR)})
    }
```

This section of code works similarly to the previous, but now reporting a degraded condition if the Deployment manifest for the nginx operand is unavailable:

```
    deployment.Namespace = req.Namespace
    deployment.Name = req.Name
    if operatorCR.Spec.Replicas != nil {
        deployment.Spec.Replicas = operatorCR.Spec.Replicas
    }
    if operatorCR.Spec.Port != nil {
        deployment.Spec.Template.Spec.Containers[0].Ports[0].
ContainerPort = *operatorCR.Spec.Port
    }
    ctrl.SetControllerReference(operatorCR, deployment,
r.Scheme)

    if create {
        err = r.Create(ctx, deployment)
    } else {
        err = r.Update(ctx, deployment)
    }
if err != nil {
    meta.SetStatusCondition(&operatorCR.Status.Conditions,
metav1.Condition{
        Type:               "OperatorDegraded",
        Status:             metav1.ConditionTrue,
        Reason:             "OperandDeploymentFailed",
        LastTransitionTime: metav1.NewTime(time.Now()),
        Message:            fmt.Sprintf("unable to update operand
deployment: %s", err.Error()),
    })
    return ctrl.Result{}, utilerrors.NewAggregate([]error{err,
r.Status().Update(ctx, operatorCR)})
}
```

If the attempt to update the operand Deployment fails, this block will report that the Operator is degraded as well. If the controller is able to succeed past this point, then there is no degradation to report. For this reason, the next block of code will finish the function by updating the Operator's CRD to show that there is no degraded condition:

```
    meta.SetStatusCondition(&operatorCR.Status.Conditions,
metav1.Condition{
        Type:                 "OperatorDegraded",
        Status:               metav1.ConditionFalse,
        Reason:               "OperatorSucceeded",
        LastTransitionTime: metav1.NewTime(time.Now()),
        Message:              "operator successfully reconciling",
    })
    return ctrl.Result{}, utilerrors.NewAggregate([]error{err,
r.Status().Update(ctx, operatorCR)})
}
```

In this code, we have added four calls to SetStatusCondition(), in which the condition with the type of "OperatorDegraded" is updated with the current time and a brief reason:

1. If the Operator is unable to access its custom resource (for any reason besides a simple IsNotFound error), set the condition to True with the reason, OperatorResourceNotAvailable.

2. If we are unable to get the nginx Deployment manifest from the embedded YAML file, update the condition to True with the reason, OperandDeploymentNotAvailable.

3. If the Deployment manifest is successfully found but a call to create or update it fails, set the condition to True with the reason, OperandDeploymentFailed.

4. Finally, if the Reconcile() function has completed with no critical errors, set the OperatorDegraded condition to False.

> **True or False as a Success Indicator**
>
> Note that, as discussed in the status conventions in *Chapter 3, Designing an Operator – CRD, API, and Target Reconciliation*, the `False` condition indicates a successful run. We could just as easily have inverted this logic, naming the condition something like `OperatorSucceeded`, where the default case is `True` and any failures change the condition to `False`. Doing so would still be consistent with Kubernetes conventions, so the decision is ultimately up to the developer based on the intent they wish to convey.

In this example, we have used string literals for each `Reason` update. In a practical application, it is common to define various `Reason` constants in the Operator's API, which allows for consistent reusability. For example, we could define the following in `api/v1alpha1/nginxoperator_types.go` and use them through their constant names:

```
const (
 ReasonCRNotAvailable        = "OperatorResourceNotAvailable"
 ReasonDeploymentNotAvailable = "OperandDeploymentNotAvailable"
  ReasonOperandDeploymentFailed = "OperandDeploymentFailed"
  ReasonSucceeded                 = "OperatorSucceeded"
 )
```

The exact naming scheme for a condition reason is up to the preference of the developer, with the only Kubernetes condition being that it must be CamelCase. For that reason, different condition types and reasons can be named whatever is relevant and preferred for that specific Operator. The only standard condition name that currently exists is `Upgradeable`, which is consumed by the OLM. We will show how to use that condition in the next section.

With these conditions implemented, we will be able to see the following output when interacting with a custom resource for our Operator:

```
$ kubectl describe nginxoperators/cluster
Name:         cluster
Namespace:    nginx-operator-system
API Version:  operator.example.com/v1alpha1
Kind:         NginxOperator
Metadata:
  Creation Timestamp:  2022-01-20T21:47:32Z
  Generation:          1
```

```
...
Spec:
  Replicas:  1
Status:
  Conditions:
    Last Transition Time:   2022-01-20T21:47:32Z
    Message:                operator successfully reconciling
    Reason:                 OperatorSucceeded
    Status:                 False
    Type:                   OperatorDegraded
```

In the next section, we will show how to report Operator conditions directly to the OLM.

Using the OLM OperatorCondition

We have already discussed how the OLM is capable of managing the currently installed list of Operators, including upgrading and downgrading. In *Chapter 7, Installing and Running Operators with the Operator Lifecycle Manager*, we will show the OLM in action for some of these features. But, for the moment, we can implement condition reporting in our Operator so that the OLM is aware of certain states that could prevent the Operator from upgrading. Combined with status conditions that are unique to the Operator (as reported in its CRD), this can help inform the OLM of critical information regarding the Operator's current state.

To read Operator conditions, the OLM creates a custom resource called `OperatorCondition`. The OLM will automatically create an instance of this object for every Operator that it manages from a CRD. A sample `OperatorCondition` object looks like the following:

```
apiVersion: operators.coreos.com/v1
kind: OperatorCondition
metadata:
  name: sample-operator
  namespace: operator-ns
status:
  conditions:
  - type: Upgradeable
    status: False
    reason: "OperatorBusy"
```

```
        message: "Operator is currently busy with a critical task"
        lastTransitionTime: "2022-01-19T12:00:00Z"
```

This object uses the same `Condition` type from the Kubernetes API as shown earlier (when implementing status conditions in the Operator CRD). This means it also includes all of the same fields, such as `type`, `status`, `reason`, and `message`, which can be updated the same way we did before. The difference is that now, setting `type` to `Upgradeable` will instruct the OLM to block any attempts to upgrade the Operator.

The other difference is that the Operator needs to report this status change to a different CRD (rather than its own). To do that, there is a library available at `https://github.com/operator-framework/operator-lib`, which includes helper functions to update the `OperatorCondition` CRD. Details on using this library are available in the Operator Framework enhancements repository at `https://github.com/operator-framework/enhancements/blob/master/enhancements/operator-conditions-lib.md`. One way this can be done is in our nginx Operator by modifying the `Reconcile()` function in `controllers/nginxoperator_controller.go` like so:

```go
import (
...

    apiv2 "github.com/operator-framework/api/pkg/operators/v2"
    "github.com/operator-framework/operator-lib/conditions"
)
func (r *NginxOperatorReconciler) Reconcile(ctx context.
Context, req ctrl.Request) (ctrl.Result, error) {
...
    condition, err := conditions.InClusterFactory{r.Client}.
    NewCondition(apiv2.ConditionType(apiv2.Upgradeable))

    if err != nil {
    return ctrl.Result{}, err
    }

    err = condition.Set(ctx, metav1.ConditionTrue,
    conditions.WithReason("OperatorUpgradeable"),
    conditions.WithMessage("The operator is upgradeable"))

    if err != nil {
```

```
    return ctrl.Result{}, err
  }
  ...
}
```

This code imports two new modules: the Operator Framework V2 API and the `operator-lib/conditions` library. It then instantiates a new `Factory` object, which uses the same Kubernetes client that is already available to the Operator. That factory can then create new `Condition` objects with `NewCondition()`, which accepts `ConditionType` (which is really just a string) and creates a condition with that type.

In this example, `Condition` is created with the `apiv2.Upgradeable` type, which is a constant defined by the Operator Framework for the `Upgradeable` condition that is understood by the OLM.

Next, the `condition.Set()` function updates the `OperatorCondition` CRD object that the OLM created for our Operator. Specifically, this adds (or updates) the list of conditions with the new condition we just created and the status that is passed to it (in this case, `True`). There are also two functions available that can be optionally passed to `Set()` (`WithReason()` and `WithMessage()`), which set the reason and message for the condition.

Using these helpers greatly simplifies the work necessary to retrieve and update the `OperatorCondition` CRD object created for our Operator by the OLM. In addition, the OLM takes steps to make sure that Operators cannot delete the CRD or modify anything outside of the object's status. However, there are some other fields in the `OperatorCondition` CRD spec that can be modified by administrators. Specifically, the `overrides` field allows users to manually bypass automatically reported condition updates that would otherwise block upgrading. A sample usage of this field looks as follows:

```
apiVersion: operators.coreos.com/v1
kind: OperatorCondition
metadata:
  name: sample-operator
  namespace: operator-ns
spec:
  overrides:
  - type: Upgradeable
    status: True
    reason: "OperatorIsStable"
```

```
    message: "Forcing an upgrade to bypass bug state"
status:
  conditions:
  - type: Upgradeable
    status: False
    reason: "OperatorBusy"
    message: "Operator is currently busy with a critical task"
    lastTransitionTime: "2022-01-19T12:00:00Z"
```

Using `overrides` like this can be useful in the case of known issues or bugs that should not prevent the Operator from upgrading.

Implementing metrics reporting

Metrics are a crucial aspect of any Kubernetes cluster. Metrics tools can provide detailed insights into almost any measurable data in the cluster. This is why metrics are a key part of graduating an Operator to Level IV in the Capability Model. In fact, most native Kubernetes controllers already report metrics about themselves, for example, **kube-scheduler** and **kube-controller-manager**. These components export data in metrics such as `schedule_attempts_total`, which reports the number of attempts the scheduler has made to schedule Pods onto Nodes.

The original design for the Kubernetes monitoring architecture (`https://github.com/kubernetes/design-proposals-archive/blob/main/instrumentation/monitoring_architecture.md`) defines metrics such as `schedule_attempts_total` as **service metrics**. The alternative to service metrics is **core metrics**, which are metrics that are generally available from all components. Core metrics currently include information about CPU and memory usage and are scraped by the Kubernetes **metrics-server** application (`https://github.com/kubernetes-sigs/metrics-server`).

On the other hand, service metrics expose application-specific data that is defined in the code of individual components. All of this data can be scraped and aggregated by tools such as **Prometheus** (`https://prometheus.io`) or **OpenTelemetry** (`https://opentelemetry.io`), and presented with frontend visualization applications such as **Grafana** (`https://grafana.com`).

The entire concept of metrics in Kubernetes extends far beyond just implementation with regard to Operators. While that, unfortunately, means there is a lot of information that is outside the scope of this chapter, there is a great deal of community and documentation resources available that cover the fundamentals of this topic. Instead, this section will focus on the relevant Operator implementation steps for a new service metric with an assumed prior understanding of the basics related to metrics in general.

Adding a custom service metric

The boilerplate code that is scaffolded by `operator-sdk` when the project is initialized already includes the code and dependencies necessary to expose a metrics endpoint in the Operator Pod. By default, this is the `/metrics` path on port `8080`. This eliminates the need for meticulous HTTP handler code and allows us to focus instead on simply implementing the metrics themselves, as described in the Prometheus documentation (`https://prometheus.io/docs/guides/go-application/#adding-your-own-metrics`).

> **Kubebuilder Metrics**
>
> The built-in metrics handler code is another aspect of the Operator SDK that is actually provided by Kubebuilder under the hood. This implementation relies on the metrics library imported from `sigs.k8s.io/controller-runtime`. That library includes features such as a global registry of metrics that is already available to the core Operator code. This library is easy to hook into in order to register new metrics and update them from anywhere in the Operator's code base. More information on metrics and their use is available in the kubebuilder book at `https://book.kubebuilder.io/reference/metrics.html`.

The `controller-runtime` library that this functionality is built with includes several metrics about the Operator's controller code already, each indicated by the prefix `controller_runtime_`. These include the following:

- `controller_runtime_reconcile_errors_total` – A counter metric that shows the cumulative number of times the `Reconcile()` function returned a non-nil error

- `controller_runtime_reconcile_time_seconds_bucket` – A histogram showing the latency for individual reconciliation attempts

- `controller_runtime_reconcile_total` – A counter that increases with every attempt to call `Reconcile()`

For this example, we will be recreating the last metric, `controller_runtime_reconcile_total`, as a way to report the number of attempts our Operator has made to reconcile its state in the cluster.

RED metrics

When it comes to what type of metrics to define in any application, the possibilities are practically infinite. This can cause overwhelming decision fatigue for developers, as it may seem there is no good place to start. So, what are the most important metrics for an Operator to expose? The Operator SDK documentation recommends following the **RED method**, which outlines three key types of metrics that every service should expose in a Kubernetes cluster:

- **Rate** – A metric showing the number of requests or attempts per second. This can provide insight into how much work an Operator is doing, highlighting frequent re-queues, which could indicate `hotloop` conditions or suboptimal request pipelines.

- **Errors** – Metrics showing the number of failed attempts by a service (for example, `controller_runtime_reconcile_errors_total`). When correlated with the Rate metric, this can help debug common failures that are degrading an Operator's performance.

- **Duration** – Duration (or latency) metrics show the length of time it takes for an Operator to complete its work (`controller_runtime_reconcile_time_seconds`). This information can indicate poor performance or other conditions that degrade the cluster's health.

These basic metrics can provide the foundation for defining **service-level objectives** (**SLOs**) to ensure your Operator functions within the expected standards of your application.

To begin adding a custom metric to our Operator, it is a good practice to organize metrics definitions into their own package by creating a new file under `controllers/metrics/metrics.go`. This new module will hold the declaration of our new metric and automatically register it with the global registry available from `sigs.k8s.io/controller-runtime/pkg/metrics`. In the following code, we define this file and instantiate a new custom metric:

controllers/metrics/metrics.go:

```go
package metrics

import (
    "github.com/prometheus/client_golang/prometheus"
    "sigs.k8s.io/controller-runtime/pkg/metrics"
)

var (
    ReconcilesTotal = prometheus.NewCounter(
        prometheus.CounterOpts{
            Name: "reconciles_total",
            Help: "Number of total reconciliation attempts",
        },
    )
)

func init() {
    metrics.Registry.MustRegister(ReconcilesTotal)
}
```

This file depends on the Prometheus client library to define a new counter (a simple, increasing metric) that we store in the `ReconcilesTotal` public variable. The actual name of the metric is `reconciles_total`, which is the name that will be exposed to Prometheus.

> **Metrics Naming Best Practices**
>
> In a real environment, it is best to include a prefix to a metric's name that is specific to the application exporting that metric. This is one of the best practices recommended by **Prometheus** (`https://prometheus.io/docs/practices/naming/`). Although we haven't done that here (for simplicity), we have followed another best practice of appending `_total` to this cumulative metric. It is helpful to be familiar with these practices, not just for developing your own metrics, but also to know what to expect when interacting with other metrics.

With this file's `init()` function automatically registering the new metric with the global registry, we can now update this metric from anywhere in the Operator's code. Since this is a measure of the total number of reconciliation attempts by the controller, it makes sense to update it at the start of the `Reconcile()` function declaration in `controllers/nginxoperator_controller.go`. This can be done with two new lines of code, shown in the following code:

controllers/nginxoperator_controller.go:

```
import (
    ...
    "github.com/sample/nginx-operator/controllers/metrics"
    ...
)
func (r *NginxOperatorReconciler) Reconcile(...) (...) {
    metrics.ReconcilesTotal.Inc()
    ...
}
```

All that is necessary is to import the new `metrics` package just created and call `metrics.ReconcilesTotal.Inc()`. This function doesn't return anything, so there is no need to add any error handling or status updates. We also want to make this the first line of the function since the goal is to increment for every call to `Reconcile()` (regardless of whether the call succeeds or not).

The updated metric value is automatically reported by the metrics endpoint initialized by kubebuilder, so it is available to view through a properly configured Prometheus instance, as shown here:

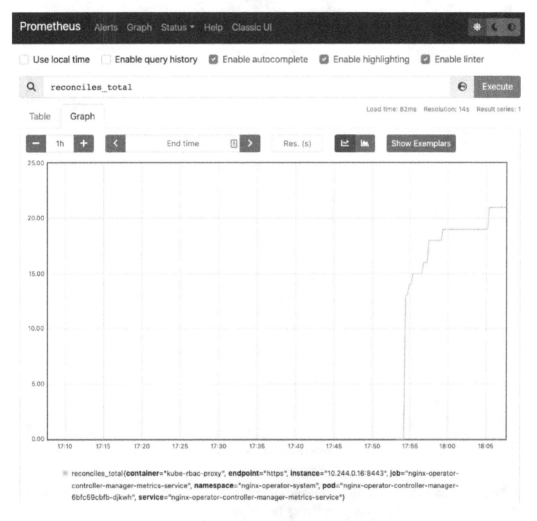

Figure 5.1 – Screenshot of reconciles_total metric graph in the Prometheus UI

When compared to the built-in `controller_runtime_reconcile_total` metric, we see that the values are the same:

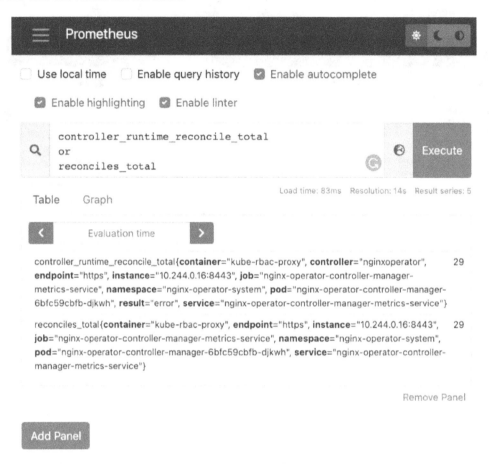

Figure 5.2 – Screenshot comparing custom reconciles_total metric to built-in controller_runtime_reconcile_total

We will cover more about how to install and configure Prometheus to capture this custom metric in *Chapter 6, Building and Deploying Your Operator*.

Implementing leader election

Leader election is an important concept in any distributed computing system, not just Kubernetes (and not just for Operators, either). High-availability applications will often deploy multiple replicas of their workload Pods to support the uptime guarantees their users expect. In situations where only one workload Pod can do work in a cluster at a time, that replica is known as the **leader**. The remaining replicas will wait, running but not doing anything significant, until the current leader becomes unavailable or gives up its status as the leader. Those Pods will then determine among themselves who should be the new leader.

Enabling proper leader election can greatly benefit the application's uptime. This can include graceful failover handling if one replica fails, or help to maintain application accessibility during rolling upgrades.

The Operator SDK makes leader election for Operators simple to implement. The boilerplate code scaffolded by operator-sdk creates a flag in our Operator, --leader-elect, which defaults to false and disables leader election. That flag is passed to the LeaderElection option value in the controller initialization function ctrl.NewManager():

main.go:

```
func main() {
...

    var enableLeaderElection bool
    flag.BoolVar(&enableLeaderElection, "leader-elect", false,
        "Enable leader election for controller manager. "+
            "Enabling this will ensure there is only one active
controller manager.")
...


    mgr, err := ctrl.NewManager(ctrl.GetConfigOrDie(), ctrl.
Options{
...

        HealthProbeBindAddress: probeAddr,
        LeaderElection:         enableLeaderElection,
        LeaderElectionID:       "df4c7b26.example.com",
    })
```

```
if err != nil {
    setupLog.Error(err, "unable to start manager")
    os.Exit(1)
}
```

The `LeaderElectionID` field represents the name of the resource that the Operator will use in order to hold the leader lock.

This setup (using the `LeaderElection` option in `ctrl.NewManager()`) sets up the Operator to use the first of two possible leader election strategies, known as **leader-with-lease**. The other possible strategy for leader election is **leader-for-life**. There are some benefits and trade-offs to each strategy, as described in the Operator SDK documentation (`https://sdk.operatorframework.io/docs/building-operators/golang/advanced-topics/#leader-election`):

- **Leader-with-lease** – The default leader election strategy wherein the current leader periodically attempts to renew its status as the leader (known as its *lease*). If it is unable to do so, the leader relinquishes its position to a new leader. This strategy improves availability by enabling fast transitions between leaders when necessary. However, this strategy makes it possible to end up with a so-called **split-brain** scenario, in which multiple replicas believe they are the leader.

- **Leader-for-life** – The leader only gives up its position when it is deleted (and by extension, its lease lock resource is deleted via garbage collection). Doing so eliminates the possibility of multiple replicas competing for leader status, but it means that recovery from failure scenarios may be bound by the kube controller manager's Pod eviction timeout (which defaults to 5 minutes).

Solving Split-Brain in Leader-with-Lease

Leader-with-lease is vulnerable to split-brain situations where two or more replicas determine inconsistent results for the current leader state. This is because the Kubernetes client that handles leader election determines the current state of the system's leader election by observing timestamps. Internally, the client looks at the elapsed time to decide whether it is time to renew the leader lease. Therefore, the clock skew rate can become an issue in some clusters.

The client library recommends a method for clock synchronization within your cluster to address this. You can also modify the `RenewDeadline` and `LeaseDuration` fields when calling `ctrl.NewManager()` to add approximate toleration for the ratio of clock skew rate between the fastest and slowest nodes. This is because `LeaseDuration` is the amount of time that current non-leaders will wait before attempting to acquire leadership, and `RenewDeadline` is the time that the control plane will wait to attempt refreshing the current leadership. Therefore, as an example, setting `LeaseDuration` to some multiple of `RenewDeadline` will add a tolerance for that same ratio of clock skew rate within the cluster.

For example, assume the current leader runs on Node A, which keeps time accurately. If the fail over backup replica is waiting on Node B, which is slow and keeps time at half the speed of Node A, then there is a clock skew rate between Nodes A and B with a ratio of 2:1.

In this scenario, if the leader has `LeaseDuration` of 1 hour and fails for some reason, then it may take 2 hours for Node B to notice that the lease has expired and attempt to acquire a new lease.

However, with `RenewDeadline` of 30 minutes, the original leader will fail to renew its lease within that time frame. This allows the replica on the slower node to act at what it believes is only 30 minutes, but the actual time has surpassed `LeaseDuration` of 1 hour. This is an obscure detail in the leader election library, but one worth noting for Operators in clusters that may be affected by this skew. There is more discussion on this topic in the original GitHub pull request to add leader election to the Kubernetes clients at `https://github.com/kubernetes/kubernetes/pull/16830`.

To make our Operator configurable so that users can decide between leader-with-lease and leader-for-life election strategies, we can leverage the existing `--leader-elect` flag to enable or disable a call to the `leader.Become()` function:

```
import (
    "github.com/operator-framework/operator-lib/leader"
    ...
)


var (
    ...
    setupLog = ctrl.Log.WithName("setup")
    )
func main() {
    ...
    var enableLeaderElection bool
    flag.BoolVar(&enableLeaderElection, "leader-elect", false,
        "Enable leader election for controller manager. "+
            "Enabling this will ensure there is only one active
controller manager.")
    ...

    if !enableLeaderElection {
        err := leader.Become(context.TODO(), "nginx-lock")
        if err != nil {
            setupLog.Error(err, "unable to acquire leader lock")
            os.Exit(1)
        }
    }

    mgr, err := ctrl.NewManager(ctrl.GetConfigOrDie(), ctrl.
Options{
        ...
        LeaderElection:          enableLeaderElection,
        LeaderElectionID:        "df4c7b26.example.com",
    })
```

Now, our Operator will always use some method of leader election. The `leader.`
`Become()` function is a blocking call that will prevent the Operator from running if it
cannot acquire the leader lock. It will attempt to do so every second. If it succeeds, it will
create the lock as a ConfigMap (in our case, that ConfigMap will be named nginx-lock).

Adding health checks

Health checks (also known as liveness and readiness probes) are a way for any Pod to
make its current functional state discoverable by other components in the cluster.
This is usually done by way of exposing an endpoint in the container (traditionally
`/healthz` for liveness checks and `/readyz` for readiness). That endpoint can be
reached by other components (such as the kubelet) to determine whether the Pod is
healthy and ready to serve requests. The topic is covered in detail in the Kubernetes
documentation at `https://kubernetes.io/docs/tasks/configure-pod-`
`container/configure-liveness-readiness-startup-probes/`.

The code initialized by the Operator SDK in `main.go` contains the `healthz` and
`readyz` check setup by default:

main.go

```
import (
  ...
  "sigs.k8s.io/controller-runtime/pkg/healthz"
)

func main() {
...
   mgr, err := ctrl.NewManager(ctrl.GetConfigOrDie(),    ctrl.
Options{
    Scheme:                scheme,
    MetricsBindAddress:    metricsAddr,
    Port:                  9443,
    HealthProbeBindAddress: probeAddr,
    LeaderElection:        enableLeaderElection,
    LeaderElectionID:      "df4c7b26.example.com",
    })
...
```

```
    if err := mgr.AddHealthzCheck("healthz", healthz.Ping); err
!= nil {
        setupLog.Error(err, "unable to set up health check")
        os.Exit(1)
    }
    if err := mgr.AddReadyzCheck("readyz", healthz.Ping); err !=
nil {
        setupLog.Error(err, "unable to set up ready check")
        os.Exit(1)
    }


    setupLog.Info("starting manager")
    if err := mgr.Start(ctrl.SetupSignalHandler()); err != nil {
        setupLog.Error(err, "problem running manager")
        os.Exit(1)
    }
}
```

This code sets up the two endpoints to begin serving immediately before the main controller is started. This makes sense because, at this point, all other startup code has already run. It wouldn't make sense to start advertising your Operator as healthy and ready to serve requests before it has even attempted to create a client connection.

The two functions, mgr.AddHealthzCheck and mgr.AddReadyzCheck, each take two parameters: a string (for the name of the check) and a function that returns the status of the check. That function must have the following signature:

```
func(*http.Request) error
```

This signature shows that the function accepts an HTTP request (because the check is serving an endpoint that will be queried by components such as the kubelet) and returns an error (if the check fails). The healthz.Ping function that the Operator is already populated with in this code is a simple **no-operation** (**no-op**) function that always returns a nil error (indicating success). This is not very insightful, but its location in the code does provide the minimum of reporting that the Operator has successfully passed most of the initialization process.

However, following the preceding function signature, it is possible to implement custom health checks. These additional checks can be added by simply calling `mgr.AddHealthzCheck` or `mgr.AddReadyzCheck` again (depending on the type of check) and passing the new function to each call. The checks are run sequentially (in no guaranteed order) when the `/healthz` or `/readyz` endpoints are queried, and if any checks fail, then an `HTTP 500` status is returned. With this, it is possible to develop your own liveness and readiness checks with more sophisticated logic that is unique to your Operator (for example, relying on dependent components to be accessible before reporting readiness).

Summary

This chapter highlighted some additional options for functionality beyond the bare minimum that was established in *Chapter 4, Developing an Operator with the Operator SDK*. This list is obviously not exhaustive of every possibility for advanced features, but it is intended to showcase some of the most common additional features added to Operators. At this point, some of the patterns for feature development should start to become clear (for example, startup and initialization code usually goes in `main.go`, while features related to core logic can fit nicely with the controller code in `nginxoperator_controller.go`, or its own package).

The work done in this chapter shows some of the steps necessary in order to graduate an Operator from lower-level functionality to higher levels in the Capability Model. For example, metrics are a key aspect of a Level IV (*Deep Insights*) Operator and, therefore, something that is expected of the highest-function Operators available to users. In addition, leader election can help establish fail over processes (helpful in reaching Level III – *Full lifecycle*). Adding functionality like this helps build a useful and feature-rich Operator that improves application performance and, by extension, user experience.

In the next chapter, we will begin compiling the code that has been built throughout *Chapter 4, Developing an Operator with the Operator SDK*, and *Chapter 5, Developing an Operator – Advanced Functionality*. We will then demonstrate how to deploy this code and run our nginx Operator in a local Kubernetes cluster (without the use of the OLM). This will be a useful development process, as bypassing the need for the OLM in a testing environment can simplify and expedite our ability to iterate and test new changes.

6

Building and Deploying Your Operator

At this point, we have written a significant amount of code to develop an **nginx Operator**, but code itself is useless unless compiled and deployed into a cluster. This chapter will demonstrate how to do just that. Specifically, the following sections will show how to use the `make` command provided by a boilerplate Operator SDK project to build a container image and manually deploy that image in a running Kubernetes cluster. In addition, this chapter will follow up those steps with guided steps for iterative development in which new changes in the Operator are compiled and pushed to the cluster. Finally, we will offer troubleshooting resources and tips for issues that may arise during this process. Those sections will be broken into the following:

- Building a container image
- Deploying in a test cluster
- Pushing and testing changes
- Troubleshooting

Note that during the course of this chapter, running the Operator in a cluster will be done manually using local build commands. This is useful for local development and testing in non-production environments because it is quick and does not rely on additional components, minimizing the time and resources required to deploy proof-of-concept test cases. In a real environment, it is better to install and manage Operators with the **Operator Lifecycle Manager**. That process will be covered in more detail in *Chapter 7, Installing and Running Operators with the Operator Lifecycle Manager*. For now, we will proceed with local deployments in a test cluster.

Technical requirements

This chapter will rely on the code from previous chapters to build a container image and deploy that image in a Kubernetes cluster. As such, the technical requirements for this chapter necessitate access to a cluster and container management tool such as Docker. However, it is not explicitly required to use the code from the previous chapters, as the commands and processes explained will work with any `operator-sdk` project. Therefore, the minimum recommended requirements for this chapter are as follows:

- An internet connection (to pull Docker base images and push built container images to a public registry).

- Access to a running Kubernetes cluster. This can be any cluster, although it is recommended to use a tool such as **Kubernetes in Docker (kind)** (`https://kind.sigs.k8s.io/`) or **minikube** (`https://minikube.sigs.k8s.io/docs/`) so that it is not costly to destroy and recreate clusters if needed.

- An up-to-date version of `kubectl` (`https://kubernetes.io/docs/tasks/tools/#kubectl`) on your machine (in order to interact with the Kubernetes cluster).

- Docker installed locally, as well as an account on either **Docker Hub** (`https://hub.docker.com/`) or another public container registry such as **Quay** (`https://quay.io/`). You can optionally use a different local container build tool such as **Buildah** (`https://buildah.io/`), but the default `Makefile` that the Operator SDK generates in a project assumes that the `docker` binary will be available locally. Therefore, additional local setup (for example, aliasing `docker` commands to `buildah`), which is not covered in this chapter, will be necessary.

This chapter introduces several of the preceding projects, some of which involve additional setup. In addition, a few of them (such as kind) are described in this tutorial only for the purposes of creating a stock test environment to follow along with. In these cases, alternative tools can be used as noted if you are more comfortable with them. For each of the technologies introduced in this chapter, resources are provided in the *Troubleshooting* section at the end of this chapter for further assistance if needed. However, the use cases with a specific technology in this chapter have been chosen to be fairly basic to guide toward the minimal risk of technical problems.

> **Note**
>
> Using a public registry without any access credentials will make your Operator image accessible to anyone on the internet. This may be fine for following a tutorial such as this, but for production images, you may wish to look more into securing your image registry (which is out of the scope of this book).

The Code in Action video for this chapter can be viewed at: `https://bit.ly/3NdVZ7s`

Building a container image

Kubernetes is a container orchestration platform, meaning that it is designed to run applications that have been built into containers. Even the core system components for Kubernetes, such as the API server and scheduler, run as containers. So, it should come as no surprise that the Operators developed for Kubernetes must also be built and deployed as containers.

For this process, a basic understanding of the fundamentals of working with containers is helpful. Fortunately, however, the Operator SDK abstracts away much of the configuration and command-line incantations to simple `Makefile` targets. These are build macros that help to automate the process of compiling binaries and container images (as well as pushing those images to a registry and deploying them in a cluster).

To see the full list of available targets provided by the Operator SDK, run the `make help` command within the project:

```
$ make help
Usage:
  make <target>
General
  help                Display this help.
Development
```

```
    manifests           Generate WebhookConfiguration, ClusterRole
and CustomResourceDefinition objects.

    generate            Generate code containing DeepCopy,
DeepCopyInto, and DeepCopyObject method implementations.

    fmt                 Run go fmt against code.

    vet                 Run go vet against code.

    test                Run tests.
Build
    build               Build manager binary.

    run                 Run a controller from your host.

    docker-build        Build docker image with the manager.

    docker-push         Push docker image with the manager.
Deployment
    install             Install CRDs into the K8s cluster specified
in ~/.kube/config.

    uninstall           Uninstall CRDs from the K8s cluster
specified in ~/.kube/config.

    deploy              Deploy controller to the K8s cluster
specified in ~/.kube/config.

    undeploy            Undeploy controller from the K8s cluster
specified in ~/.kube/config.

    controller-gen      Download controller-gen locally if
necessary.

    kustomize           Download kustomize locally if necessary.

    envtest             Download envtest-setup locally if necessary.

    bundle              Generate bundle manifests and metadata, then
validate generated files.

    bundle-build        Build the bundle image.

    bundle-push         Push the bundle image.

    opm                 Download opm locally if necessary.

    catalog-build       Build a catalog image.

    catalog-push        Push a catalog image.
```

Some of these commands, such as make manifests and make generate, were used in earlier chapters to initialize the project and generate the Operator's API and **CustomResourceDefinition** (CRD). Now, we are more concerned with the commands under the Build heading. Specifically, make build and make docker-build, with the former responsible for compiling a local binary of the Operator and the latter building a container image.

Building the Operator locally

First, let's examine `make build`. From `Makefile`, the definition is simple:

```
build: generate fmt vet ## Build manager binary.
   go build -o bin/manager main.go
```

This target is primarily concerned with running `go build`, which is the standard command to compile any **Go** program. Also worth noting is the fact that the first things this target depends on are the `make generate`, `make fmt`, and `make vet` targets, which ensure that the Operator's generated API code is up to date and that the Go code in the project's source code conforms to the stylistic standards of the language. This is an added convenience and is why `Makefile` targets such as this are useful in development.

Running `make build` produces the standard output that one would expect when compiling a Go program:

```
$ make build
/home/nginx-operator/bin/controller-gen
object:headerFile="hack/boilerplate.go.txt" paths="./..."
go fmt ./...
go vet ./...
go build -o bin/manager main.go
```

When the compilation is successful, there should be no more output than the preceding code. Upon completion, there will now be a new executable file under `bin/manager`, which is the compiled Operator code. This can be run manually (or with `make run`) against any accessible Kubernetes cluster, but it will not actually be deployed in the cluster until it is built into a container image. This is what `make docker-build` does.

Building the Operator image with Docker

The definition for `make docker-build` is slightly more interesting than the local `build` target:

```
docker-build: test ## Build docker image with the manager.
   docker build -t ${IMG} .
```

This is essentially just calling `docker build` (with an added dependency to make a `test` that runs any unit tests defined in the project along with ensuring all generated CRD manifests and API code are up to date).

The `docker build` command will instruct the local Docker daemon to construct a container image from the Dockerfile in the root of the Operator's project directory. This file was originally generated by the initial `operator-sdk init` command from when the project was first created. We have made a very slight modification (which will be explained here), so the file now looks like the following:

Dockerfile:

```
# Build the manager binary
FROM golang:1.17 as builder
WORKDIR /workspace
# Copy the Go Modules manifests
COPY go.mod go.mod
COPY go.sum go.sum
# cache deps before building and copying source so that we
don't need to re-download as much
# and so that source changes don't invalidate our downloaded
layer
RUN go mod download
# Copy the go source
COPY main.go main.go
COPY api/ api/
COPY controllers/ controllers/
COPY assets/ assets
# Build
RUN CGO_ENABLED=0 GOOS=linux GOARCH=amd64 go build -a -o
manager main.go
# Use distroless as minimal base image to package the manager
binary
# Refer to https://github.com/GoogleContainerTools/distroless
for more details
FROM gcr.io/distroless/static:nonroot
WORKDIR /
COPY --from=builder /workspace/manager .
USER 65532:65532
ENTRYPOINT ["/manager"]
```

Note that the exact details about how this Dockerfile works are a more advanced topic in regard to container builds. It's not critical to understand each of these in depth (just one benefit of using the Operator SDK to generate the file!), but we will summarize them here. These steps roughly break down into the following:

1. Set the base image for the Operator to build with Go 1.17.

2. Copy the Go module dependency files to the new image.

3. Download the module dependencies.

4. Copy the main Operator code, including `main.go`, `api/`, `controllers/`, and `assets/` to the new image.

> **Note**
>
> In this project, we have modified the default Dockerfile to copy the `assets/` directory. When it is generated by `operator-sdk`, this Dockerfile only copies `main.go` and the `api/` and `controllers/` directories by default. Since the tutorial for our nginx Operator included adding a new top-level package under `assets/`, we need to ensure that this directory is also included in the Operator image. This serves as an example to demonstrate that it is okay to modify the project's default Dockerfile (however, using version control or otherwise making backups is recommended).
>
> Alternatively, the `assets` package could have been created under the `controllers/` folder, which would not have required any updates to the Dockerfile (because it would have been included under the existing COPY `controllers/ controllers/` line). See the *Troubleshooting* section of this chapter for more information.

5. Compile the Operator binary within the image. This is the same as building the Operator locally (as shown previously), except it will be packaged within a container.

6. Define the Operator's binary as the main command for the built container.

With the preceding Dockerfile (including the change to include COPY `assets/ assets/`), running `make docker-build` will successfully complete. But, before we do that, first note that this command includes a variable that we have not yet discussed: `${IMG}`.

The Makefile command uses this IMG variable to define the tag for the compiled image. That variable is defined earlier in Makefile with a default value of controller:latest:

```
# Image URL to use all building/pushing image targets
IMG ?= controller:latest
```

This is helpful to know because, without updating this variable, the built image for our Operator will simply have the name controller. In order to build an image with a tag that references our actual container registry (for example, docker.io/myregistry) the build command can be invoked like so:

```
$ IMG=docker.io/sample/nginx-operator:v0.1 make docker-build
...
docker build -t docker.io/sample/nginx-operator:v0.1 .
[+] Building 99.1s (18/18) FINISHED
 => [internal] load build definition from Dockerfile 0.0s
 => [builder  1/10] FROM docker.io/library/golang:1.17 21.0s
 => [builder  2/10] WORKDIR /workspace              2.3s
 => [builder  3/10] COPY go.mod go.mod              0.0s
 => [builder  4/10] COPY go.sum go.sum              0.0s
 => [builder  5/10] RUN go mod download             31.3s
 => [builder  6/10] COPY main.go main.go            0.0s
 => [builder  7/10] COPY api/ api/                  0.0s
 => [builder  8/10] COPY controllers/ controllers/ 0.0s
 => [builder  9/10] COPY assets/ assets            0.0s
 => [builder 10/10] RUN CGO_ENABLED=0 GOOS=linux GOARCH=amd64
go build -a -o manager main.go                     42.5s
 => [stage-1 2/3] COPY --from=builder /workspace/manager .
 => exporting to image                             0.2s
 => => exporting layers                            0.2s
 => => writing image sha256:dd6546d...b5ba118bdba4 0.0s
 => => naming to docker.io/sample/nginx-operator:v0.1
```

Some output has been omitted, but the important parts to note are the builder steps, which have been included. These follow the steps as defined in the project's Dockerfile.

With a container image successfully built, the new image should now be present on your local machine. You can confirm this by running `docker images`:

```
$ docker images
REPOSITORY              TAG        IMAGE ID       CREATED
SIZE
sample/nginx-operator   v0.1       dd6546d2afb0   45 hours ago
48.9MB
```

In the next section, we will push this image to a public registry and deploy the Operator in a running Kubernetes cluster.

Deploying in a test cluster

Now that the Operator has been built into a container image, it can be deployed in a cluster as a container. To do this, you will first need to ensure that you have access to a running cluster as well as a public image registry. To host your image in a registry, you can obtain a free personal account on **Docker Hub** (`https://hub.docker.com`).

For this tutorial, we will be using a local Kubernetes cluster created with kind, which deploys a running Kubernetes cluster within Docker containers rather than directly on the local machine, and is available at `https://kind.sigs.k8s.io/`. However, the steps described here will be agnostic to any Kubernetes cluster running the latest version of the platform. For example, if you are more comfortable using development environments such as minikube (or have another cluster already available), then you can skip the kind setup shown in this section. The rest of the steps in this section will apply to any Kubernetes cluster.

To start a local cluster with kind, ensure that you have Docker and kind installed on your machine and run `kind create cluster`:

```
$ kind create cluster
Creating cluster "kind" ...
 Ensuring node image (kindest/node:v1.21.1)
 Preparing nodes
 Writing configuration
 Starting control-plane
 Installing CNI
 Installing StorageClass
 Set kubectl context to "kind-kind"
You can now use your cluster with
```

```
kubectl cluster-info --context kind-kind
Have a nice day!
```

Note that `kind create cluster` may take a moment to complete. This bootstraps a functional Kubernetes cluster running within Docker. You can confirm that your cluster is accessible by running any `kubectl` command, for example, `kubectl cluster-info`:

```
$ kubectl cluster-info
Kubernetes master is running at https://127.0.0.1:56384
CoreDNS is running at https://127.0.0.1:56384/api/v1/
namespaces/kube-system/services/kube-dns:dns/proxy
```

With a cluster running, it's time to make the Operator's image accessible by pushing it to a public registry. First, ensure that you have access to your registry. For Docker Hub, this means running `docker login` and entering your username and password.

Once logged in, you can push the image using the provided `Makefile` `make docker-push` target (which is simply the equivalent of manually running `docker push`):

```
$ IMG=docker.io/sample/nginx-operator:v0.1 make docker-push
docker push docker.io/sample/nginx-operator:v0.1
The push refers to repository [docker.io/sample/nginx-operator]
18452d09c8a6: Pushed 5b1fa8e3e100: Layer already exists
v0.1: digest: sha256:5315a7092bd7d5af1dbb454c05c15c54131
bd3ab78809ad1f3816f05dd467930 size: 739
```

(This command may take a moment to run, and your exact output may differ.)

Note that we have still passed the `IMG` variable to this command. To eliminate the need to do this, you can either modify `Makefile` to change the default definition of the variable (this definition was shown in the *Building a container image* section earlier) or export your image name as an environment variable, like so:

```
$ export IMG=docker.io/sample/nginx-operator:v0.1
$ make docker-push
docker push docker.io/sample/nginx-operator:v0.1
The push refers to repository [docker.io/sample/nginx-operator]
18452d09c8a6: Pushed 5b1fa8e3e100: Layer already exists
v0.1: digest: sha256:5315a7092bd7d5af1dbb454c05c15c54131bd
3ab78809ad1f3816f05dd467930 size: 739
```

Now, the image is available publicly on the internet. You may wish to manually confirm that your image is accessible by running `docker pull <image>`, but this is not required.

Avoiding Docker Hub

You technically do not have to use a public registry such as Docker Hub to make your image accessible to the cluster. There are alternative ways of importing your image into the cluster, for example, kind provides the `kind load docker-image` command, which manually loads the image into your cluster's internal registry (see `https://kind.sigs.k8s.io/docs/user/quick-start/#loading-an-image-into-your-cluster` for more information). However, in this tutorial, we have chosen the public registry route as it is a common approach (especially for open source Operators that are published for others to use) and remains agnostic to the specific cluster you may be running.

With the Operator image accessible (and the public image name defined in an environment variable or modified in `Makefile`, as shown earlier), all that is required to run the Operator in a cluster now is the `make deploy` command:

```
$ make deploy
/Users/sample/nginx-operator/bin/controller-gen
rbac:roleName=manager-role crd webhook paths="./..."
output:crd:artifacts:config=config/crd/bases
cd config/manager && /Users/sample/nginx-operator/bin/kustomize
edit set image controller=docker.io/sample/nginx-operator:v0.1
/Users/sample/nginx-operator/bin/kustomize build config/default
| kubectl apply -f -
namespace/nginx-operator-system created
customresourcedefinition.apiextensions.k8s.io/nginxoperators.
operator.example.com created
serviceaccount/nginx-operator-controller-manager created
role.rbac.authorization.k8s.io/nginx-operator-leader-election-
role created
clusterrole.rbac.authorization.k8s.io/nginx-operator-manager-
role created
clusterrole.rbac.authorization.k8s.io/nginx-operator-metrics-
reader created
clusterrole.rbac.authorization.k8s.io/nginx-operator-proxy-role
created
rolebinding.rbac.authorization.k8s.io/nginx-operator-leader-
```

```
election-rolebinding created
clusterrolebinding.rbac.authorization.k8s.io/nginx-operator-
manager-rolebinding created
clusterrolebinding.rbac.authorization.k8s.io/nginx-operator-
proxy-rolebinding created
configmap/nginx-operator-manager-config created
service/nginx-operator-controller-manager-metrics-service
created
deployment.apps/nginx-operator-controller-manager created
```

Now, you will see a new namespace in your cluster that matches the Operator's name:

```
$ kubectl get namespaces
NAME                    STATUS    AGE
default                 Active    31m
kube-node-lease         Active    31m
kube-public             Active    31m
kube-system             Active    31m
local-path-storage      Active    31m
nginx-operator-system   Active    54s
```

Exploring this namespace deeper with kubectl get all will show that it contains a **Deployment**, **ReplicaSet**, **Service**, and **Pod** for the Operator (some output has been omitted for brevity):

```
$ kubectl get all -n nginx-operator-system
NAME                                                              READY
STATUS      pod/nginx-operator-controller-manager-6f5f66795d-
945pb    2/2      Running
NAME
TYPE         service/nginx-operator-controller-manager-metrics-
service    ClusterIP
NAME                                                              READY
UP-TO-DATE  deployment.apps/nginx-operator-controller-manager
1/1       1
NAME
DESIRED     replicaset.apps/nginx-operator-controller-manager-
6f5f66795d   1
```

But where is the operand nginx Pod? Recall that we designed the Operator to do nothing if it cannot locate an instance of its CRD. To remedy this, you can create your first CRD object (matching the API defined in *Chapter 4*, *Developing an Operator with the Operator SDK*) like the following:

sample-cr.yaml:

```
apiVersion: operator.example.com/v1alpha1
kind: NginxOperator
metadata:
  name: cluster
  namespace: nginx-operator-system
spec:
  replicas: 1
```

Create this file and save it anywhere on your machine as any name (in this case, `sample-cr.yaml` is fine). Then, run `kubectl create -f sample-cr.yaml` to create the custom resource object in the nginx Operator's namespace. Now, running `kubectl get pods` will show that the new nginx Pod (named `cluster-xxx`) has been created:

```
$ kubectl get pods -n nginx-operator-system
NAME                                                    READY
STATUS      cluster-7855777498-rcwdj
1/1      Running
nginx-operator-controller-manager-6f5f66795d-hzb8n      2/2
Running
```

You can modify the custom resource object you just created with `kubectl edit nginxoperators/cluster -n nginx-operator-system`. This command (`kubectl edit`) will open your local text editor where you can make changes directly to the object's `spec` fields. For example, to change the number of operand replicas, we can run the preceding command and set `spec.replicas: 2`, like so:

$ kubectl edit nginxoperators/cluster -n nginx-operator-system

```
apiVersion: operator.example.com/v1alpha1
kind: NginxOperator
metadata:
  creationTimestamp: "2022-02-05T18:28:47Z"
```

```
    generation: 1
    name: cluster
    namespace: nginx-operator-system
    resourceVersion: "7116"
    uid: 66994aa7-e81b-4b18-8404-2976be3db1a7
  spec:
    replicas: 2
  status:
    conditions:
    - lastTransitionTime: "2022-02-05T18:28:47Z"
      message: operator successfully reconciling
      reason: OperatorSucceeded
      status: "False"
      type: OperatorDegraded
```

Notice that when using `kubectl edit`, other fields in the Operator (such as `status`) are also visible. While you cannot directly modify these fields, this is a good spot to point out that our Operator conditions are successfully reporting in the CRD's `status` section. This is indicated by the `OperatorDegraded: False` condition type and status.

However, take note that this condition may initially be confusing to users because it appears to be indicating that `OperatorSucceeded` is `False` at first glance. But, upon further inspection, it is indicated that `OperatorSucceeded` is actually the reason for `OperatorDegraded` to be `False`. In other words, the *Operator* is *not* degraded because the Operator *succeeded*. This example has intentionally been chosen to highlight the care that must be taken to implement informative and clearly understandable status conditions.

Saving the changes to the CRD object and running `kubectl get` pods again now shows a new nginx Pod:

```
$ kubectl get pods -n nginx-operator-system
NAME                                                         READY
STATUS      cluster-7855777498-rcwdj
1/1         Running
cluster-7855777498-kzs25                                     1/1
Running
nginx-operator-controller-manager-6f5f66795d-hzb8n           2/2
Running
```

Similarly, changing the `spec.replicas` field to 0 will delete all of the nginx Pods:

```
$ kubectl get pods -n nginx-operator-system
NAME                                               READY
STATUS     cluster-7855777498-rcwdj
1/1        Terminating
cluster-7855777498-kzs25                           1/1
Terminating
nginx-operator-controller-manager-6f5f66795d-hzb8n  2/2
Running
```

This concludes the basic deployment steps for an Operator. The following steps summarize what we have done so far:

1. Built a container image for the Operator

2. Pushed the image to a public registry

3. Used `make deploy` to launch the Operator in a local cluster (in the process, pulling the image from the public registry into the cluster)

4. Manually created an instance of the Operator's CRD object

5. Modified the CRD object within the cluster using `kubectl edit`

However, there is still some work to be done in order to enable metrics (which was a big part of the work done in *Chapter 5, Developing an Operator – Advanced Functionality*, and key to achieving higher-level functionality within the **capability model**). In the next section, we will demonstrate how to make changes to our Operator's deployment and redeploy it in the cluster.

Pushing and testing changes

During the course of development (for any project, not just Kubernetes Operators) it will likely become necessary to make changes to the code or other project files (such as resource manifests) and test those changes. In the case of this example, we will not be changing any code. Instead, we will redeploy the Operator with the proper metrics resources created to make the metrics visible, which we implemented in *Chapter 5, Developing an Operator – Advanced Functionality*.

Installing and configuring kube-prometheus

Metrics are not very useful without a tool to scrape and present them. This is what Prometheus is for, and it understands the metrics language in which we have implemented our own metrics. There are a number of other tools that can parse Prometheus metrics. In this tutorial, we will use **kube-prometheus** (`https://github.com/prometheus-operator/kube-prometheus`) to install a full end-to-end monitoring stack in our cluster. kube-prometheus provides a number of additional features that we won't explicitly explore in this book, but it is a very convenient and powerful library for installing monitoring in a cluster. In your own environment, you may choose another option, such as installing Prometheus directly or using the Prometheus Operator from `https://github.com/prometheus-operator/prometheus-operator` (which is provided by kube-prometheus).

To get started, follow the steps at `https://github.com/prometheus-operator/kube-prometheus` installing to install kube-prometheus in our Operator project. Take note of the prerequisites for the *Installing* and *Compiling* sections in that link. Specifically, the following tools are required:

- `jb`
- `gojsontoyaml`
- `jsonnet`

When kube-prometheus has been successfully installed in the project, we will have a new subdirectory (`my-kube-prometheus`, as described in the kube-prometheus documentation), which contains the following files:

```
$ ls
total 20K
drwxr-xr-x  8 ... .
drwxr-xr-x 21 ... ..
drwxr-xr-x 74 ... manifests
drwxr-xr-x 19 ... vendor
-rwxr-xr-x  1 ... build.sh
-rw-r--r--  1 ... 05 example.jsonnet
-rw-r--r--  1 ... jsonnetfile.json
-rw-r--r--  1 ... 04 jsonnetfile.lock.json
```

Now, we will modify `example.jsonnet` to include our Operator's namespace. This means modifying the `values+::` block within the file to add a `prometheus+` object that includes a list of namespaces (in our case, only the `nginx-operator-system` namespace):

my-kube-prometheus/example.jsonnet:

```
local kp =
  (import 'kube-prometheus/main.libsonnet') +
  ...
  {
    values+:: {
      common+: {
        namespace: 'monitoring',
      },

      prometheus+: {
        namespaces+: ['nginx-operator-system'],
      },
    },
  };
...
```

Next, use `build.sh` to compile the new manifests by running the following command:

```
$ ./build.sh example.jsonnet
```

Now, we can create the `kube-prometheus` manifests in our cluster by applying them with the following commands (from within the `my-kube-prometheus` directory):

```
$ kubectl apply --server-side -f manifests/setup
$ kubectl apply -f manifests/
```

When finished, the Prometheus dashboard should be accessible by running the following commands to open a local proxy to your cluster and the Prometheus service:

```
$ kubectl --namespace monitoring port-forward svc/prometheus-k8s 9090
Forwarding from 127.0.0.1:9090 -> 9090
Forwarding from [::1]:9090 -> 9090
```

(Note, this command will remain running until you manually end it, for example, by pressing *Ctrl + C*.) The dashboard will be visible by navigating to `http://localhost:9090` in your web browser. However, if you try to search for our Operator's metric (recall that it was named `reconciles_total`), you will see that it is not available. This is because we need to redeploy our Operator with an additional manifest that is not created by default.

Redeploying the Operator with metrics

Prometheus knows to scrape our Operator's namespace for metrics due to the configuration created previously. However, it still needs to know which specific endpoint within the namespace to query. This is the role of an object called `ServiceMonitor` (`https://pkg.go.dev/github.com/coreos/prometheus-operator/pkg/apis/monitoring/v1#ServiceMonitor`). This object is not created by default by the Operator SDK when running `make deploy`, so we need to modify `config/default/kustomization.yaml`. (This file location is relative to the project root directory, not the new `my-kube-prometheus` directory that we created previously when installing kube-prometheus).

In this file, simply find any lines that are marked with `[PROMETHEUS]` and uncomment them by removing the leading pound or the hash symbol (#). This is currently only one line, shown in the following:

config/default/kustomization.yaml:

```
...
bases:
- ../crd
- ../rbac
- ../manager
# ...
# ...
# [PROMETHEUS] To enable prometheus monitor,
uncomment all sections with 'PROMETHEUS'.
- ../prometheus
```

This is the default configuration file for **Kustomize** (`https://kustomize.io/`), which is a Kubernetes templating project that the Operator SDK uses to generate and deploy project manifests.

At this point, you can run `make undeploy` to remove the current Operator installation, followed by running `make deploy` once again to recreate it. After a few moments, the `reconciles_total` metric should be visible in the Prometheus dashboard. The following screenshot shows this metric in the Prometheus dashboard search bar:

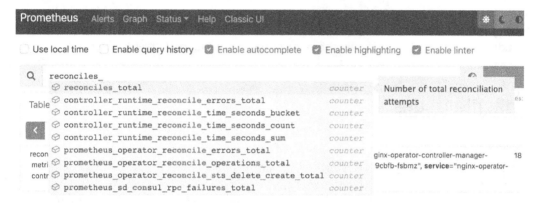

Figure 6.1 – Screenshot of the Prometheus dashboard

Key takeaways

While this section may seem focused on just setting up metrics, it actually covers some important steps related to the development-driven redeployment of an Operator project. Specifically, we covered the following:

- Installing `kube-prometheus` as a library in our project and configuring it to scrape our Operator's namespace

- Modifying Kustomize configuration files to include new dependencies in our Operator deployment

- Using `make undeploy` to remove the existing Operator deployment

Technically, we could have simply run `make deploy` without first undeploying the project. The idempotent nature of Kubernetes resource manifests means that only the new resources would have been created. However, awareness of `make undeploy` is very useful in cases where the existing project may need to be completely removed.

Troubleshooting

This chapter introduced several new tools and concepts not yet covered by earlier chapters. These include Docker, kind, kubectl, Make, and kube-prometheus. It is possible that you may have run into some issues while working with these tools, so this section is intended to offer links to references that can help resolve common issues. Many of the underlying tools used in this chapter are not exclusive to the Operator Framework, which thankfully means that there is a wealth of resources available to address problems you may encounter.

Makefile issues

Make (`https://www.gnu.org/software/make/`) is a very popular tool for automating the generation and compilation of countless software projects. It was already used in *Chapter 4, Developing an Operator with the Operator SDK,* and *Chapter 5, Developing an Operator – Advanced Functionality,* in order to generate the APIs and manifests used by our project. In this chapter, it was leveraged even more to automate many of the commands for building and deploying.

However, the `make ...` commands used throughout this book are shorthand for running underlying tools. Therefore, when encountering an error with any `make` commands, the first debugging step is to examine the `Makefile` to find what that command is actually running. If this happens, you will likely find that you are rather encountering an issue with Docker, Go, or Kubernetes.

These commands have been preemptively provided when `operator-sdk` initialized the project, but you are free to modify the provided `Makefile` as you wish to customize your project.

kind

In this chapter, we used kind to deploy a test Kubernetes cluster. Using kind offers a very quick way to create (and destroy) local Kubernetes clusters. It also provides a configurable setup that allows for relatively easy changes to the default cluster (for example, starting a cluster with additional nodes).

The official website for kind is `https://kind.sigs.k8s.io/`. The website provides extensive documentation and sample configurations for different cluster setups. In addition, the kind code base is available on GitHub at `https://github.com/kubernetes-sigs/kind`. The kind maintainers and users are also active on the official Kubernetes Slack server (`slack.k8s.io`) in the **#kind** channel. Both of these links are excellent resources for asking questions or searching for answers.

Docker

If you are working with Kubernetes, you are likely already familiar with Docker
(https://www.docker.com/). It is just one of several options for building and
managing container images, which are essential for deploying applications on Kubernetes.
The key step for transitioning Operator code to a deployable image is the docker build
command (which is called when running make docker-build). This command
follows the build steps defined in the Dockerfile. More information on Dockerfile syntax
is available in the Docker documentation at https://docs.docker.com/engine/
reference/builder/.

When building a container image following the steps in this chapter, there are some
unique issues specific to this tutorial that you may encounter, which are explained next.

docker build fails with no required module error for assets package

When running make docker-build, you may find that your build fails with the
following error (or something similar):

```
$ make docker-build

...

=> ERROR [builder 9/9] RUN CGO_ENABLED=0 GOOS=linux
GOARCH=amd64 go build -a -o manager main.go
2.1s

------

 > [builder 9/9] RUN CGO_ENABLED=0 GOOS=linux GOARCH=amd64 go
build -a -o manager main.go:
#15 2.081 controllers/nginxoperator_controller.go:37:2: no
required module provides package github.com/sample/nginx-
operator/assets; to add it:
#15 2.081   go get github.com/sample/nginx-operator/assets

------

executor failed running [/bin/sh -c CGO_ENABLED=0 GOOS=linux
GOARCH=amd64 go build -a -o manager main.go]: exit code: 1
make: *** [docker-build] Error 1
```

This error is actually arising from the go build command in the final step of the
Dockerfile. In the context of this specific tutorial, Go is failing to build because it cannot
locate the assets package (created in *Chapter 4, Developing an Operator with the
Operator SDK*, to organize and access the operand Deployment manifest).

To fix this, ensure that you have modified the Dockerfile to include the `COPY assets/ assets/` directive (see the example in the *Building a container image* section of this chapter). Alternatively, you could refactor the Operator code to embed the `assets/` directory within the existing `controllers/` directory without needing to modify the Dockerfile. This is because both `controllers/` and `api/` are already copied to the `builder` image (but it does not semantically make sense to store embedded manifests within the API directory, as they are not an API).

docker push fails with access denied

The `make docker-push` command may fail with the following error:

```
$ make docker-push
docker push controller:latest
The push refers to repository [docker.io/library/controller]
18452d09c8a6: Preparing
5b1fa8e3e100: Preparing
denied: requested access to the resource is denied
make: *** [docker-push] Error 1
```

This exact error (including the `docker push controller:latest` line) implies a misconfigured `IMG` variable for the command. Recall that in the *Building a container image* section of this chapter, this variable was discussed as a way to tag the built image with an appropriate name. There are a few options for setting this value, either as an environment variable or by modifying the `Makefile`.

However you choose to update this variable, it is important to check that the value is propagated to the `docker-push` target in the `Makefile` as well. Otherwise, Docker will attempt to push this to a generic registry for library images. You do not have access to this registry, therefore, Docker returns the `access denied` error shown here.

If instead, the error does include the appropriate public registry with your `IMG` variable value (for example, the second line is `docker push docker.io/yourregistry/yourimagename`) then it is likely a simple authentication error. Run `docker login` to make sure you are logged into your Docker Hub account.

Operator deploys but fails to start with ImagePull error

If you run make deploy, the command will likely always complete successfully (unless you have made significant modifications to the generated manifest files). However, when viewing the Operator in your cluster (for example, with kubectl get all -n nginx-operator-system) you may see that the Operator's Pod is failing to start with the following message:

```
$ kubectl get all -n nginx-operator-system
NAME                     READY    STATUS             RESTARTS    AGE
pod/nginx-operator...    1/2      ImagePullBackOff   0           34s
```

This is probably a similar error to the one described previously. In Kubernetes, the ImagePullBackOff error indicates that, for some reason, the Pod is unable to pull the container image it is intended to run. This is usually either an authentication error (for example, the registry may be private) or the image is not found. Ensure that you have built and pushed the Operator image with the right IMG environment variable set, as mentioned in the other *Troubleshooting* sections. If you still see the error, check to make sure that your image registry is not set to private by logging into the Docker Hub web UI.

Operator deploys but fails to start with another error

In Kubernetes, there are many reasons why any given Pod may fail to start. There could be a logical bug in the Operator's code or there may be a system configuration issue within the cluster. There is, unfortunately, no *one-size-fits-all* solution to this problem. However, there are tools to gather more information. Using kubectl to inspect the Pod is the most common way to diagnose errors. For example, kubectl describe pod/<podname> will print events and status updates that can explain the failure. Or, kubectl logs pod/<podname> will print the log output of the Pod (which is helpful for diagnosing runtime errors that usually need to be fixed in code). All of the kubectl commands will provide documentation by running kubectl -h.

Metrics

As part of developing an Operator with rich features and debugging capabilities, this and previous chapters dedicated effort to demonstrating the implementation of Prometheus metrics. Prometheus (https://prometheus.io/) is an open source monitoring platform that is widely used in the Kubernetes ecosystem. Therefore, there are many resources available online for various problems (many are not specific to the Operator Framework). These community resources are documented at https://prometheus.io/community/.

Operator metrics do not show up in Prometheus

With Prometheus deployed in a cluster following the kube-prometheus steps in this tutorial, the custom metrics defined in our Operator code should show up in the Prometheus dashboard after a few moments. If, after some time, the custom metrics are still not visible, ensure that you have made the correct changes described in the *Installing and configuring kube-prometheus* section to instruct Prometheus to scrape the Operator's namespace for new metrics.

Additionally, make sure that you have uncommented the - ../prometheus line in config/default/kustomization.yaml before deploying the Operator as described in the *Redeploying the Operator with metrics* section. This step ensures that the ServiceMonitor object (which informs Prometheus which endpoint to scrape for metrics) is created in the namespace.

Operator deployment fails with no matches for kind 'ServiceMonitor'

When running make deploy, the following error may appear among several other lines showing which resources were created:

```
$ make deploy
...
serviceaccount/nginx-operator-controller-manager created
role.rbac.authorization.k8s.io/nginx-operator-leader-election-
role created
clusterrole.rbac.authorization.k8s.io/nginx-operator-manager-
role created
clusterrole.rbac.authorization.k8s.io/nginx-operator-metrics-
reader created
clusterrole.rbac.authorization.k8s.io/nginx-operator-proxy-role
created
...
error: unable to recognize "STDIN": no matches for kind
"ServiceMonitor" in version "monitoring.coreos.com/v1"
make: *** [deploy] Error 1
```

In this case, the Operator did not actually fail to deploy. However, it did fail to create the `ServiceMonitor` object due to an inability to locate the object's definition in the Kubernetes API. This is likely due to failing to install Prometheus in the cluster before attempting to deploy the Operator with metrics. Specifically, `ServiceMonitor` is a CRD that is provided by Prometheus. So, deploying the Operator with metrics before installing `kube-prometheus` in this tutorial will lead to failures when reporting metrics. To resolve this, ensure that you have followed the steps for installing Prometheus before deploying the Operator with metrics.

Additional errors

The preceding issues are just some of the technical problems that may arise when following this tutorial. It is, unfortunately, not possible for this chapter to cover every scenario. However, the Operator Framework community and its resources provide solutions to many different types of problems. These, along with the resources in the *Troubleshooting* section of *Chapter 4, Developing an Operator with the Operator SDK*, are very likely to resolve any difficulty that you may face.

Summary

This chapter's main objective was to compile the Operator code that we have been building throughout this book and deploy it in a cluster. To do this, we followed steps designed for local development environments. These included building the Operator as a local binary and building a container image to deploy in an ephemeral test cluster (created using kind). This lightweight process is helpful for development and rapid testing, but it lacks the full workflow benefits needed for publishing an Operator with the intent of deploying in production.

In the next chapter, we will explore the final pillars of the Operator Framework: **OperatorHub** and the **Operator Lifecycle Manager**. Learning how to prepare and submit an Operator to OperatorHub will be a key part of offering any Operator available for public use. With that, the Operator Lifecycle Manager is a much more elegant solution for deploying Operators (both publicly available on OperatorHub or privately deployed). Compared to deploying manually with Make, these processes are much better suited for sharing your Operator with the world.

Part 3: Deploying and Distributing Operators for Public Use

This section will show you how to package your Operator and publish it for others to use, contributing to the open source ecosystem that is essential to the Operator Framework with the help of case studies. Finally, the lessons from this book will be summarized in an FAQ-style breakdown before applying these lessons to real-world examples of application and system Operators.

This section comprises the following chapters:

- *Chapter 7, Installing and Running Operators with the Operator Lifecycle Manager*
- *Chapter 8, Preparing for Ongoing Maintenance of Your Operator*
- *Chapter 9, Diving into FAQs and Future Trends*
- *Chapter 10, Case Study for Optional Operators – the Prometheus Operator*
- *Chapter 11, Case Study for Core Operator – Etcd Operator*

7

Installing and Running Operators with the Operator Lifecycle Manager

Up until now, the Operator development work covered in previous chapters has been mostly self-contained. That is, the development and deployment processes covered so far have been primarily focused on local environments with relatively few external services expected to interact with the Operator we have been writing. While these processes are useful (and in some ways essential) to the early design of an Operator, there is an expectation for most Operators (and indeed, most software projects in general) that they will eventually be exposed to the outside world. This chapter will focus on this phase of an Operator's lifespan, wherein the Operator is presented and consumed by external users.

In *Chapter 1, Introduction to the Operator Framework*, the three main pillars of the Operator Framework were introduced. Several chapters of this book have already been devoted to the first pillar (the Operator SDK), but the remaining pillars have yet to be explored in detail. These are the **Operator Lifecycle Manager (OLM)** and **OperatorHub**. These two components of the Operator Framework are the key transitory elements in an Operator's development from an experimental, local prototype to a published, installable product. In this chapter, we will cover the necessary steps to graduate from an Operator in development to one that is accessible by users, through the following sections:

- Understanding the OLM
- Running your Operator
- Working with OperatorHub
- Troubleshooting

By packaging an Operator to be installed and managed by the OLM and then publishing that Operator on OperatorHub, we will be leveraging the standard deployment workflow that users expect from the Operator Framework. These steps are by no means necessary, as we have already shown that it is possible to manually build and deploy an Operator without the OLM or OperatorHub. But, it is the goal of this chapter to introduce these pillars of the Operator Framework to demonstrate how we can transform an Operator into a rich community project.

Technical requirements

This chapter will continue to work with the nginx Operator that was written in *Chapter 4, Developing an Operator with the Operator SDK,* and *Chapter 5, Developing an Operator – Advanced Functionality*. It will also assume access to a public Docker registry (previously used in *Chapter 6, Building and Deploying Your Operator*), as well as access to a running **Kubernetes** cluster. Therefore, the technical requirements of this chapter build upon most of the requirements from previous chapters, including the following:

- Access to a Kubernetes cluster. It is recommended to use a disposable cluster created with a tool such as **kind** or **minikube** (see *Chapter 6, Building and Deploying Your Operator*).
- The `kubectl` binary available on your local system for interacting with the Kubernetes cluster.
- The `operator-sdk` binary available on your local system for deploying the OLM and building Operator manifests.

- Docker installed and running to build Operator bundle images.
- A GitHub account and familiarity with the GitHub fork and pull request processes for submitting a new Operator to OperatorHub (demonstration only).

The Code in Action video for this chapter can be viewed at: `https://bit.ly/3PPItsB`

Understanding the OLM

The OLM was introduced in *Chapter 1, Introduction to the Operator Framework,* as a tool for installing and managing Operators within a cluster. Its features include the ability to provide control over upgrading installed Operators and making these Operators visible to cluster users. It also helps maintain cluster stability by enforcing Operator dependencies and preventing conflicting APIs from different Operators. This is a brief overview, but these features make it a powerful tool for deploying Operators in production environments. You can find more details about the OLM's features in the Operator Framework documentation at `https://olm.operatorframework.io/docs/#features-provided-by-olm`.

While this may make the OLM seem like a complex component to work with, it is actually no more than a set of resource manifests that can be installed in a cluster similarly to any other component or application (including Operators themselves). These resources include various **Pods** (managed by Deployments), **CustomResourceDefinitions** (**CRDs**), **namespaces**, **ServiceAccounts**, and **RoleBindings**.

In addition, the Operator SDK command-line tools provide simple commands for easily installing and interacting with the OLM in a Kubernetes cluster.

So, before it is possible to install Operators with the OLM, we must first install the OLM itself. This section will show the steps required to do so. It will also demonstrate some additional commands for interacting with the OLM via the command line, which will be helpful later on when installing and managing our own Operator.

Installing the OLM in a Kubernetes cluster

To install the OLM, first ensure that you have administrative access to a running Kubernetes cluster. Even though using the OLM to manage Operators is an acceptable practice for production clusters, it is strongly recommended to use a disposable cluster (created with a tool such as kind) while following along with this chapter. This makes it easy and affordable to destroy and re-build the cluster from scratch if necessary. If you already have a cluster running from a previous chapter, it may even be useful to shut down that cluster in order to start fresh (with kind, the command to do so is `kind delete cluster`).

Next, invoke the `operator-sdk` binary to install the OLM in your cluster with the following command:

```
$ operator-sdk olm install
INFO[0000] Fetching CRDs for version "latest"
INFO[0000] Fetching resources for resolved version "latest"
...
INFO[0027]  Deployment "olm/packageserver" successfully rolled
out INFO[0028] Successfully installed OLM version "latest"
```

This command may take a moment to complete, but during that time you will see `operator-sdk` fetching the various resource manifests for the OLM and installing them in your Kubernetes cluster. Once this is complete, it will also print the final list of installed resources. Many of these are either cluster-scoped (such as the OLM-specific CRDs) or installed in the newly created `olm` namespace. You can see these resources by inspecting that namespace with `kubectl` using the following command:

```
$ kubectl get all -n olm
NAME                                       READY     STATUS
RESTARTS    AGE
pod/catalog-operator-5c4997c789-xr986      1/1       Running     0
4m35s
pod/olm-operator-6d46969488-nsrcl          1/1       Running     0
4m35s
pod/operatorhubio-catalog-h97sx            1/1       Running     0
4m27s
pod/packageserver-69649dc65b-qppvg         1/1       Running     0
4m26s
pod/packageserver-69649dc65b-xc2fr         1/1       Running     0
4m26s
NAME                            TYPE          CLUSTER-IP
EXTERNAL-IP
service/operatorhubio-catalog   ClusterIP     10.96.253.116
<none>          service/packageserver-service    ClusterIP
10.96.208.29     <none>
NAME                            READY     UP-TO-DATE
AVAILABLE    AGE
deployment.apps/catalog-operator    1/1     1                  1
4m35s
```

```
deployment.apps/olm-operator          1/1     1            1
4m35s
deployment.apps/packageserver         2/2     2            2
4m26s


NAME                                              DESIRED    CURRENT
READY       replicaset.apps/catalog-operator-5c4997c789    1
1           1            replicaset.apps/olm-operator-6d46969488
1           1        1            replicaset.apps/packageserver-
69649dc65b       2          2          2
```

Notably, there are five Pods in this namespace that perform the core functions of the OLM. These Pods work together to provide the cohesive functionality that comprises the OLM, including tracking Operator subscriptions and watching for custom resources that indicate Operator installations in the cluster.

Interacting with the OLM

Along with `operator-sdk olm install` (which, as the name implies, installs the OLM in a cluster), the `operator-sdk` binary also provides two more OLM-specific commands: `olm uninstall` and `olm status`. The former will remove the OLM and all of its dependent manifests from your cluster, while the latter provides information on the current status of the OLM resources in the cluster. For a healthy OLM installation, that output looks like this:

```
$ operator-sdk olm status
INFO[0000] Fetching CRDs for version "v0.20.0"
INFO[0000] Fetching resources for resolved version "v0.20.0"
INFO[0002] Successfully got OLM status for version "v0.20.0"
NAME                                              NAMESPACE
KIND                      STATUS
operatorgroups.operators.coreos.com
CustomResourceDefinition      Installed
operatorconditions.operators.coreos.com
CustomResourceDefinition      Installed
olmconfigs.operators.coreos.com
CustomResourceDefinition      Installed
installplans.operators.coreos.com
CustomResourceDefinition      Installed
clusterserviceversions.operators.coreos.com
CustomResourceDefinition      Installed
```

```
olm-operator-binding-olm
ClusterRoleBinding          Installed

operatorhubio-catalog                           olm
CatalogSource               Installed

olm-operators                                   olm
OperatorGroup               Installed

aggregate-olm-view
ClusterRole                 Installed

catalog-operator                                olm
Deployment                  Installed

cluster
OLMConfig                   Installed

operators.operators.coreos.com
CustomResourceDefinition    Installed

olm-operator                                    olm
Deployment                  Installed

subscriptions.operators.coreos.com
CustomResourceDefinition    Installed

aggregate-olm-edit
ClusterRole                 Installed

olm
Namespace                   Installed

global-operators                          operators
OperatorGroup               Installed

operators
Namespace                   Installed

packageserver                                   olm
ClusterServiceVersion       Installed

olm-operator-serviceaccount                     olm
ServiceAccount              Installed

catalogsources.operators.coreos.com
CustomResourceDefinition    Installed

system:controller:operator-lifecycle-manager
ClusterRole                 Installed
```

However, if the OLM was not behaving properly or there were issues with Operators in your cluster, this command could be used to debug the cause. For example, you can run `kubectl delete crd/operatorgroups.operators.coreos.com` (which deletes the `OperatorGroups` CRD installed by the OLM). Following this, running `operator-sdk olm status` will show the error `no matches for kind "OperatorGroup" in version "operators.coreos.com/v1` next to the `global-operators` and `olm-operators` entries, indicating that the CRD is missing in the cluster.

This error can be repaired by uninstalling the OLM with `operator-sdk olm uninstall` and reinstalling it. Note that uninstalling the OLM does not uninstall any of the Operators it manages in the cluster. This is intentional to prevent data loss, but it also means that any desire to remove Operators from the cluster cannot be done by simply uninstalling the OLM.

Besides installing and checking on the health of the OLM itself, the other way to interact with it is by installing and managing Operators. But first, the Operator must be prepared in a way that the OLM will understand. This is called **the Operator's bundle**, and we will show how to generate it in the next section.

Running your Operator

In *Chapter 6, Building and Deploying Your Operator*, we demonstrated ways to build and run an Operator manually by either compiling locally or building a Docker image to run in a Kubernetes cluster. But, neither of these methods is directly compatible with the OLM, so in order to provide an Operator that can be installed by the OLM, the Operator must be prepared with a bundle that contains metadata about the Operator in a format that the OLM understands. Then, this bundle can be passed to the OLM, which will handle the rest of the installation and life cycle management of the Operator.

Generating an Operator's bundle

An Operator's bundle consists of various manifests that describe the Operator and provide additional metadata, such as its dependencies and APIs. Once created, these manifests can be compiled into a **bundle image**, which is a deployable container image that is used by the OLM to install the Operator in a cluster.

The simplest way to generate the bundle manifests is by running `make bundle`. This command will ask you to provide some metadata about the Operator and compile that input into the output resource manifests.

> **Note**
>
> `make bundle` generates a container image name in some fields based on the same IMG environment variable used in *Chapter 6, Building and Deploying Your Operator*. Ensure that this environment variable is still set when generating the bundle, or that it is otherwise being passed to the `make bundle` command when it is invoked.

The following block shows the output of `make bundle`. In this case, we will fill out the prompts for our nginx Operator with the company name, `MyCompany`, as well as some additional keywords and contact information for the maintainers of the Operator:

```
$ make bundle
/Users/mdame/nginx-operator/bin/controller-gen
rbac:roleName=manager-role crd webhook paths="./..."
output:crd:artifacts:config=config/crd/bases
operator-sdk generate kustomize manifests -q
Display name for the operator (required):
> Nginx Operator
Description for the operator (required):
> Operator for managing a basic Nginx deployment
Provider's name for the operator (required):
> MyCompany
Any relevant URL for the provider name (optional):
> http://mycompany.example
Comma-separated list of keywords for your operator (required):
> nginx,tutorial
Comma-separated list of maintainers and their emails (e.g.
'name1:email1, name2:email2') (required):
> Mike Dame:mike@mycompany.example
cd config/manager && /Users/mdame/nginx-operator/bin/kustomize
edit set image controller=controller:latest
/Users/mdame/nginx-operator/bin/kustomize build config/
manifests | operator-sdk generate bundle -q --overwrite
--version 0.0.1
INFO[0000] Creating bundle.Dockerfile
INFO[0000] Creating bundle/metadata/annotations.yaml
INFO[0000] Bundle metadata generated suceessfully
operator-sdk bundle validate ./bundle
INFO[0000] All validation tests have completed successfully
```

During this step, the generator will request the following inputs one by one:

- `Display name for the operator`: This is the name that will be used for displaying the Operator on resources such as OperatorHub. So, it should be readable and clear with proper capitalization. For example, we have chosen `Nginx Operator`.

- `Description for the operator`: This field provides a description of the Operator and its functionality. Similar to the display name, this is intended for users to see. Therefore, it should also be clear and thorough to describe the Operator's functionality in detail.

- `Provider's name for the operator`: This is the name of the provider, or developer, of the Operator. For a single developer, it can simply be your name. Or, for larger organizations, it could be a company or department name.

- `Any relevant URL for the provider name`: This is the opportunity for developers to provide an external URL to find more information about the developer. This could be a personal blog, GitHub account, or corporate website.

- `Comma-separated list of keywords for your operator`: This is a list of keywords that can help users categorize and find your Operator. For this example, we have chosen `nginx,tutorial`, but you could just as easily provide a different list, such as `deployment,nginx,high availability,metrics`. This gives more insight into the key functionality we have developed for this Operator. Note also that the list is comma-separated, so `high availability` is one keyword.

- `Comma-separated list of maintainers and their emails`: Finally, this section is a chance to provide the contact information for the maintainers of the Operator. This gives users information on who to contact for support or bug reporting. However, it can be useful for the developer's privacy to provide a corporate address rather than personal contact information.

These fields correspond to matching fields in the Operator's **cluster service version (CSV)** file (the CSV was briefly described in *Chapter 1, Introduction to the Operator Framework*, and will be explained in more detail later in this chapter under *Working with OperatorHub*). You can find more information about how each of these fields is used in the Operator Framework documentation at `https://sdk.operatorframework.io/docs/olm-integration/generation/#csv-fields`.

The CSV is one of several new files created in the project after running `make bundle`. The majority of these new files are created under a new directory called `bundle/`. There is also a new file at the root of the project called `bundle.Dockerfile`, which is used to compile the manifests into the bundle image.

Exploring the bundle files

The files generated by `make bundle` contain metadata about the Operator that can be used by the OLM to install and manage the Operator, as well as OperatorHub, to provide information to users about the Operator and its dependencies and capabilities. Within the `bundle/` directory are three subdirectories that contain the following files:

- `tests/`: These are configuration files for running scorecard tests, which are a series of tests designed to validate the Operator's bundle (see `https://sdk.operatorframework.io/docs/testing-operators/scorecard`).

- `metadata/`: This contains an `annotations.yaml` file, which provides the OLM with information about an Operator's version and dependencies. The annotations in this file must be the same as the labels specified in `bundle.Dockerfile` (more on that file shortly), and should usually not be modified.

- `manifests/`: This directory contains various manifests required by your operator, including the Operator's CRD and metrics-related resources (if applicable). Most notably, however, is the CSV, which contains the bulk of the Operator's metadata.

The Operator's CSV is the most interesting of these files, as it contains much of the information used by the OLM to process the creation of the Operator, as well as OperatorHub, to display important information to users about the Operator. The one created for our nginx Operator is named `nginx-operator.clusterserviceversion.yaml`, and contains the following sections:

1. Metadata, including a sample custom resource object (to be created by the user for configuring the Operator) and its capability level:

```
apiVersion: operators.coreos.com/v1alpha1
kind: ClusterServiceVersion
metadata:
  annotations:
    alm-examples: |-
      [
        {
```

```json
      "apiVersion": "operator.example.com/v1alpha1",
      "kind": "NginxOperator",
      "metadata": {
        "name": "nginxoperator-sample"
      },

      "spec": null
    }

  ]
```

```yaml
  capabilities: Basic Install
  operators.operatorframework.io/builder: operator-
sdk-v1.17.0
  operators.operatorframework.io/project_layout:
go.kubebuilder.io/v3
 name: nginx-operator.v0.0.1
 namespace: placeholder
```

2. A specification field with the Operator's description, display name, display icon (if provided), and related CRDs:

```yaml
spec:
  apiservicedefinitions: {}

  customresourcedefinitions:
    owned:
    - description: NginxOperator is the Schema for the
nginxoperators API
      displayName: Nginx Operator
      kind: NginxOperator
      name: nginxoperators.operator.example.com
      version: v1alpha1
  description: Operator for managing a basic Nginx
deployment
  displayName: Nginx Operator
  icon:
  - base64data: ""
    mediatype: ""
```

3. Installation instructions, including the cluster permissions and Deployment specification for the Operator Pod (omitted here for brevity).

4. The install modes for the Operator, showing which namespace installation strategies it supports:

```
installModes:
- supported: false
  type: OwnNamespace
- supported: false
  type: SingleNamespace
- supported: false
  type: MultiNamespace
- supported: true
  type: AllNamespaces
```

5. Keywords, maintainer information, provider URL, and version (as provided when running make bundle):

```
keywords:
- nginx
- tutorial
links:
- name: Nginx Operator
  url: https://nginx-operator.domain
maintainers:
- email: mike@mycompany.example
  name: Mike Dame
maturity: alpha
provider:
  name: MyCompany
  url: http://mycompany.example
version: 0.0.1
```

Together, this information can be packaged together to provide enough data for the OLM to deploy and manage the Operator in a cluster. That package is known as the bundle image.

Building a bundle image

Once the bundle manifests have been generated, the bundle image can be built by calling make bundle-build. This command builds a Docker container based on the bundle.Dockerfile file that was generated earlier by make bundle. That Dockerfile file contains the following instructions:

```
$ cat bundle.Dockerfile
FROM scratch
# Core bundle labels.
LABEL operators.operatorframework.io.bundle.mediatype.
v1=registry+v1
LABEL operators.operatorframework.io.bundle.manifests.
v1=manifests/
LABEL operators.operatorframework.io.bundle.metadata.
v1=metadata/
LABEL operators.operatorframework.io.bundle.package.v1=nginx-
operator
LABEL operators.operatorframework.io.bundle.channels.v1=alpha
LABEL operators.operatorframework.io.metrics.builder=operator-
sdk-v1.17.0
LABEL operators.operatorframework.io.metrics.mediatype.
v1=metrics+v1
LABEL operators.operatorframework.io.metrics.project_layout=go.
kubebuilder.io/v3
# Labels for testing.
LABEL operators.operatorframework.io.test.mediatype.
v1=scorecard+v1
LABEL operators.operatorframework.io.test.config.v1=tests/
scorecard/
# Copy files to locations specified by labels.
COPY bundle/manifests /manifests/
COPY bundle/metadata /metadata/
COPY bundle/tests/scorecard /tests/scorecard/
```

Similar to the main Dockerfile file used to compile the Operator image in *Chapter 6, Building and Deploying Your Operator*, one of the key steps in this Dockerfile's build is to copy the essential bundle files from the bundle/ directory into its own image (highlighted in the preceding code block). It also labels the resulting image with various metadata about the operator, its versions, and the tools used to build it.

Running `make bundle-build` produces the following build log:

```
$ make bundle-build
docker build -f bundle.Dockerfile -t example.com/nginx-
operator-bundle:v0.0.1 .
[+] Building 0.4s (7/7) FINISHED

 => [internal] load build definition from bundle.Dockerfile
 => => transferring dockerfile: 966B
 => [internal] load .dockerignore

 => => transferring context: 35B
 => [internal] load build context

 => => transferring context: 16.73kB
 => [1/3] COPY bundle/manifests /manifests/
 => [2/3] COPY bundle/metadata /metadata/
 => [3/3] COPY bundle/tests/scorecard /tests/scorecard/
 => exporting to image

 => => exporting layers

 => => writing image
sha256:6b4bf32edd5d15461d112aa746a9fd4154fefdb1f9cfc49b
56be52548ac66921
 => => naming to example.com/nginx-operator-bundle:v0.0.1
```

However, note that the name of the new container image is example.com/nginx-operator-bundle, which you can confirm by running docker images:

```
$ docker images
REPOSITORY                                 TAG        IMAGE ID
CREATED            example.com/nginx-operator-bundle    v0.0.1
6b4bf32edd      29 seconds ago
```

This generic name is used because make bundle-build depends on a different environment variable than the IMG variable that was used earlier to build the Operator image manually (and generate the bundle manifests). To set a custom bundle image name, either tag the generated image or re-run make bundle-build with the BUNDLE_IMG variable set. An example is shown here:

```
$ BUNDLE_IMG=docker.io/myregistry/nginx-bundle:v0.0.1 make
bundle-build
```

This will generate the bundle image with the name docker.io/myregistry/nginx-bundle:v0.0.1.

Pushing a bundle image

Recall that in *Chapter 6, Building and Deploying Your Operator*, it was necessary to not only build the container image for the Operator but also push it to a publicly accessible registry. This made the image available to our Kubernetes cluster. Similarly, the bundle image must also be accessible by the cluster (and the OLM). For this reason, we must also push the bundle image to a registry so that the OLM can pull it into the cluster.

The Operator SDK makes this step easy with the make bundle-push command:

```
$ make bundle-push
/Library/Developer/CommandLineTools/usr/bin/make docker-push
IMG=docker.io/mdame/nginx-bundle:v0.0.1
docker push docker.io/mdame/nginx-bundle:v0.0.1
The push refers to repository [docker.io/mdame/nginx-bundle]
79c3f933fff3: Pushed
93e60c892495: Pushed
dd3276fbf1b2: Pushed
v0.0.1: digest: sha256:f6938300b1b8b5a2ce127273e2e48443
ad3ef2e558cbcf260d9b03dd00d2f230 size: 939
```

This command simply calls docker push, but it inherits the environment variables that have been set and used in previous commands (for example, BUNDLE_IMG). This convenience helps reduce the chance of making a mistake and pushing the wrong image name to the wrong registry.

Deploying an Operator bundle with the OLM

With a bundle image built and pushed to an accessible registry, it is simple to deploy the Operator from its bundle with the `operator-sdk run bundle` command. For example, we can now deploy the nginx Operator bundle from the previous section by running the following:

```
$ operator-sdk run bundle docker.io/mdame/nginx-bundle:v0.0.1
INFO[0013] Successfully created registry pod: docker-io-mdame-
nginx-bundle-v0-0-1
INFO[0013] Created CatalogSource: nginx-operator-catalog
INFO[0013] OperatorGroup "operator-sdk-og" created
INFO[0013] Created Subscription: nginx-operator-v0-0-1-sub
INFO[0016] Approved InstallPlan install-44bh9 for the
Subscription: nginx-operator-v0-0-1-sub
INFO[0016] Waiting for ClusterServiceVersion "default/nginx-
operator.v0.0.1" to reach 'Succeeded' phase
INFO[0016]   Waiting for ClusterServiceVersion "default/nginx-
operator.v0.0.1" to appear
INFO[0023]   Found ClusterServiceVersion "default/nginx-
operator.v0.0.1" phase: Pending
INFO[0026]   Found ClusterServiceVersion "default/nginx-
operator.v0.0.1" phase: Installing
INFO[0046]   Found ClusterServiceVersion "default/nginx-
operator.v0.0.1" phase: Succeeded
INFO[0046] OLM has successfully installed "nginx-operator.
v0.0.1"
```

> **Note**
>
> This command may take a few minutes to succeed. However, if the Operator's `ClusterServiceVersion` object fails to install, double-check that you have followed the steps to install **kube-prometheus** in the cluster, as detailed in *Chapter 6, Building and Deploying Your Operator*. If the Operator bundle has been built to include references to Prometheus resources, and these resources are not present in the cluster, this can cause the Operator's installation to fail.

This command creates the resources necessary for the OLM to install the nginx Operator using only the information in the Operator's bundle, including `CatalogSource`, `OperatorGroup`, `Subscription`, and `InstallPlan`.

The Operator can then be uninstalled using the `operator-sdk cleanup <packageName>` command, where `<packageName>` is defined in the Operator's `PROJECT` file as `projectName`:

```
$ operator-sdk cleanup nginx-operator
INFO[0000] subscription "nginx-operator-v0-0-1-sub" deleted
INFO[0000] customresourcedefinition "nginxoperators.operator.
example.com" deleted
INFO[0000] clusterserviceversion "nginx-operator.v0.0.1"
deleted
INFO[0000] catalogsource "nginx-operator-catalog" deleted
INFO[0000] operatorgroup "operator-sdk-og" deleted
INFO[0000] Operator "nginx-operator" uninstalled
```

This concludes the normal development workflow for building and deploying Operators with the OLM manually. However, there is another source for Operators to install in a cluster. This is **OperatorHub**, and it is the focus of the next section.

Working with OperatorHub

Any successful open source project requires a dedicated community of users and developers to help the project's ecosystem thrive. The Operator Framework is no different, and at the center of this community is the Operator catalog of `https://operatorhub.io/`. In fact, this centralization is the exact goal of OperatorHub, as stated on `https://operatorhub.io/about`:

> *While there are several approaches to implement Operators yielding the same level of integration with Kubernetes, what has been missing is a central location to find the wide array of great Operators that have been built by the community. OperatorHub.io aims to be that central location.*

Launched in 2019 by a collaborative effort between **Red Hat**, **AWS**, **Google Cloud**, and **Microsoft**, OperatorHub has been a driving force in the growth and adoption of Kubernetes Operators. As of the time of writing, the OperatorHub index contains over 200 Operators (and this number continues to grow). Backed by only a public GitHub repository and many volunteer maintainers, the open concept of catalog moderation and acceptance of OperatorHub supports the very ideals of Kubernetes, by allowing anyone from any organization to contribute their Operator to the catalog and make it accessible to all.

In short, OperatorHub makes it easy to promote your own Operators, as well as finding and installing Operators developed by other providers. In this section, we'll demonstrate how to do both of these by working with the OperatorHub website and backend.

Installing Operators from OperatorHub

Installing an Operator from OperatorHub in your own Kubernetes cluster is very easy using the catalog on `https://operatorhub.io/`. You can begin by browsing the list of all available Operators, or by searching in the text box on the OperatorHub home page. You can also narrow down your search by category (available categories include **AI/Machine Learning**, **Big Data**, **Cloud Provider**, and **Monitoring**).

For an arbitrary example, the **Grafana** Operator can be found under **Monitoring**. Grafana is an analytics and monitoring visualization platform that provides rich, insightful tools for viewing application health. The following is a screenshot showing **Grafana Operator** and others available in the **Monitoring** category on OperatorHub:

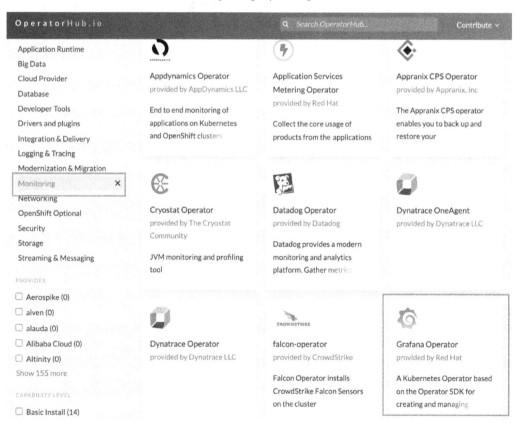

Figure 7.1 – Screenshot of the OperatorHub Monitoring category

Clicking on the **Grafana Operator** tile opens up the information page for this specific Operator. This page includes information such as the Operator's current functionality level in the Capability Model, which versions of the Operator have been published, and information about the provider and maintainer of the Operator. The following is a screenshot showing what the **Grafana Operator** information page looks like on OperatorHub:

Figure 7.2 – Grafana Operator information page

Also available on this page are installation instructions for this Operator, found by clicking the **Install** button on the top right of the Operator's description. This opens a new window with commands that can be copied and pasted to install the OLM, the Operator itself, and how to watch the Operator startup. Since we already have the OLM installed in a cluster, we can skip to *Step 2* and copy the install command for this Operator. This is shown in the following screenshot of the Grafana Operator installation instructions:

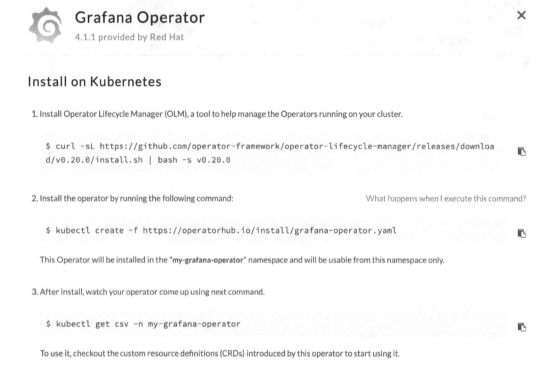

Figure 7.3 – Grafana Operator installation instructions

Running this command in a terminal produces the following output:

```
$ kubectl create -f https://operatorhub.io/install/grafana-
operator.yaml
namespace/my-grafana-operator created
operatorgroup.operators.coreos.com/operatorgroup created
subscription.operators.coreos.com/my-grafana-operator created
```

Following this, the new namespace, my-grafana-operator, has been created with the resources necessary for this Operator:

```
$ kubectl get all -n my-grafana-operator
NAME                                                          READY
STATUS    pod/grafana-operator-controller-manager-b95954bdd-
sqwzr    2/2      Running
NAME
TYPE          service/grafana-operator-controller-manager-
metrics-service    ClusterIP
NAME                                                          READY
UP-TO-DATE    deployment.apps/grafana-operator-controller-
manager    1/1      1                1
NAME
DESIRED    replicaset.apps/grafana-operator-controller-manager-
b95954bdd    1            1
```

In addition, this command also created an OperatorGroup object and Subscription object for this Operator. These resource types are CRDs that were installed in the cluster by the OLM and implemented by individual Operators to represent their installation. The details about what each of these objects does are available in the OperatorHub documentation at https://operatorhub.io/how-to-install-an-operator, but in summary, they define the user's (your) intent to install the Operator and inform the OLM about the location of the Operator's metadata on OperatorHub. The OLM uses this information to create the Deployment, Service, and other resources needed for the new Operator.

Once an Operator has been installed, it is usually up to the user to create the configuration CRD object for that Operator. With so many different CRDs floating around, this can get confusing to keep track of. However, many of these CRDs (such as OperatorGroup and Subscription) are installed and managed automatically by tools such as the OLM, and they do not require manual interaction. Generally, the user is only concerned with the CRD object for a specific Operator's configuration (such as the one that we created for our nginx Operator). Most Operator README files and OperatorHub descriptions will contain example CRDs and steps to get started with each Operator (and it is a good practice to do so with your own Operator as well).

Speaking of your own Operator, contributing to the OperatorHub catalog is almost as easy as installing Operators from it. In the next section, we'll look at how each of these Operators made their way onto OperatorHub and how yours can, too.

Submitting your own Operator to OperatorHub

While it is not required to publish any Operator publicly, many providers choose to do so, both for the benefit of the community and their own users. If the Operator you have developed is used to manage an application you offer to your users, public availability of the Operator can increase awareness of the application and bolster your organization's reputation among the open source community. Offering a free Operator shows your users that you are invested in providing a stable product with minimal engineering hours required on their part.

> **What Is Required from Operator SDK Projects?**
>
> The Operator SDK, like many Kubernetes projects, is released under the **Apache 2.0 License**. This gives permissive usability of the project for commercial use, distribution, and private use (among other use cases). More information on the Operator SDK license is available at `https://github.com/operator-framework/operator-sdk/blob/master/LICENSE`.

Because the nginx Operator that we have been developing throughout this book is only intended as a tutorial (and not meant for public use), we cannot demonstrate the process for submitting it to OperatorHub. However, the general process for submitting an Operator to OperatorHub is outlined at `https://operatorhub.io/contribute`. In broad terms, this involves the following steps:

1. Develop an Operator that is ready for publishing.

2. Generate the Operator's bundle, including its CSV and related CRDs.

3. Create a **pull request** (**PR**) against the OperatorHub repository on GitHub (`https://github.com/k8s-operatorhub/community-operators`) with your Operator's metadata.

If you have been following the steps in this book up until this point, then you are already familiar with the first two steps. However, the third step is the most important part of submitting to OperatorHub, because the GitHub repository represents the entire catalog of Operators listed on OperatorHub. So, without the proper PR changes necessary to merge your Operator's information into this repository, it will not show up on OperatorHub.

> **Which OperatorHub Repository is which?**
>
> Some outdated documentation that is still available refers to two different OperatorHub repository locations, `community-operators` and `upstream-community-operators`, which were originally subdirectories of the now-archived OperatorHub repository at `https://github.com/operator-framework/community-operators`. The former is a remnant of the initial work done by Red Hat to publish OperatorHub (specifically, it refers to a location that was reserved for Operators to be listed on an integrated version of OperatorHub within Red Hat's OpenShift distribution of Kubernetes). This OpenShift-specific Operator index has since been decoupled from the community repository referenced earlier. There is documentation available for contributing to the OpenShift catalog for developers who are interested in doing so, but this chapter will focus on the community OperatorHub, which is platform-agnostic.

The steps to submit your Operator through GitHub are as follows (these steps assume some prior familiarity with GitHub and the fork/PR processes involved):

1. Fork the OperatorHub repository (`https://github.com/k8s-operatorhub/community-operators`) into your own GitHub account. This allows you to clone a local copy of the repository to your machine and make changes to it that will later be pulled into the upstream catalog via your PR.

2. Create a new folder for your Operator under the `operators/` directory. It must have a unique name from all other Operators (for example, we could create `operators/nginx-operator`).

3. Create a new file called `ci.yaml` in this directory. This file defines versioning semantics as well as the reviewers allowed to make changes to your Operator (more information available at `https://k8s-operatorhub.github.io/community-operators/operator-ci-yaml/`). A simple `ci.yaml` file looks like the following:

    ```
    reviewers:
      - myGithubUsername
      - yourTeammateUsername
    ```

4. Create a directory in your Operator's folder for each version you wish to publish (for example, `operators/nginx-operator/0.0.1`).

5. Copy the contents of the `bundle` directory from your Operator's project into the new version folder.

Also, copy the `bundle.Dockerfile` that was generated at your Operator's project root into the version folder.

6. Commit and push the changes to a new branch of your forked OperatorHub repository on GitHub.

7. Navigate back to the upstream OperatorHub repository's PR page (`https://github.com/k8s-operatorhub/community-operators/pulls`) and click **New pull request**. Choose your fork and branch to merge into the upstream repository.

8. Read the PR template description and ensure that you have followed all of the steps outlined. These prerequisite steps help expedite the review and approval process of your Operator's PR and include the following:

    I.   Reviewing the community contribution guidelines

    II.   Testing your Operator in a local cluster

    III. Verifying your Operator's metadata aligns with the standards of OperatorHub

    IV. Ensuring your Operator's description and versioning schema are sufficient for your users

Once you have reviewed the pre-submission checks in the PR template, submit your request. At this point, automated checks will validate your Operator's metadata to ensure it passes the quality thresholds for submission (and report any problems in a GitHub comment). If you need to make any changes to your submission in order for it to pass these checks, you can simply push the changes to your forked branch of the OperatorHub repository.

Once your PR passes the pre-submission checks, it should automatically merge your changes into the upstream repository. Soon thereafter, your Operator will be visible on `https://operatorhub.io/` for the world to install!

Troubleshooting

While this chapter introduced some new concepts, including the OLM and OperatorHub, many of the resources already listed in earlier *Troubleshooting* sections throughout this book still apply.

OLM support

The OLM and its related resources are related in general to Operator SDK development. So, there is reasonable help for this topic available in the `#operator-sdk-dev` Slack channel on `slack.k8s.io`. The OLM is also available on GitHub for reporting issues at `https://github.com/operator-framework/operator-lifecycle-manager`. The documentation for integrating an Operator with the OLM is available as a primary resource at `https://sdk.operatorframework.io/docs/olm-integration/`.

OperatorHub support

OperatorHub is also available on GitHub at the catalog repository shown in this chapter (`https://github.com/k8s-operatorhub/community-operators`). For issues with the frontend `https://operatorhub.io/` website specifically, that code is located at `https://github.com/k8s-operatorhub/operatorhub.io`. This repository provides detailed documentation on all of the necessary metadata and bundle files for OperatorHub submission (as well as the submission process itself) at `https://k8s-operatorhub.github.io/community-operators/`.

OperatorHub also provides validators and tools for previewing your Operator's submission prior to creating a PR against the repository. The preview tool is available at `https://operatorhub.io/preview`. Submitting the generated CSV in this tool will show a preview of how your Operator will look once it is submitted to OperatorHub:

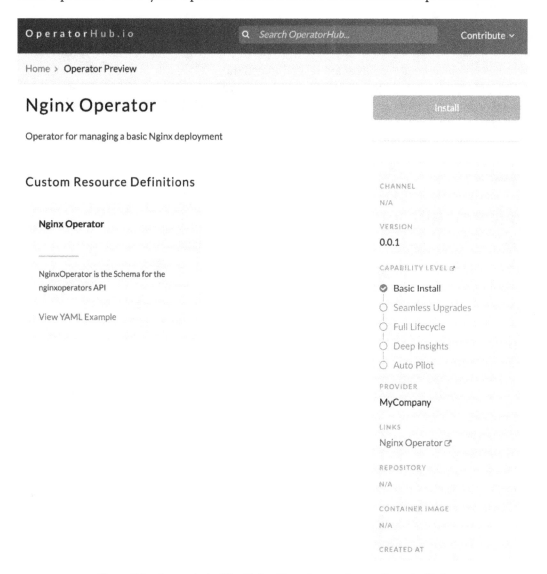

Figure 7.4 – Screenshot of the Nginx Operator preview on OperatorHub

Previewing an Operator's presentation can be a very helpful manual step in testing that all of the metadata prepared for that Operator is going to present to new users in the way that you want. It can be easy to lose track of the confusing CRD and CSV definitions, so previewing it gives early visual confirmation that everything is set up correctly. It also verifies that the metadata is syntactically valid.

Summary

This chapter concluded the primary development and publication of an Operator. If you have been following along up until this point while developing your own Operator, then congratulations! Your Operator is now published and accessible to new users thanks to the reach of OperatorHub. Starting from the early chapters of this book, we have shown the steps to design an Operator, develop its basic and advanced functionality with Go, build and deploy it for local testing, and finally, package and publish it for public distribution. However, very few Operator project life cycles will end at this point.

It is likely that most Operators will eventually need to evolve, change their provided features, and release new versions. Doing so in a consistent and predictable way benefits both your users and your maintainers by establishing expected release standards. These standards include policies for deprecation and timelines for new releases. In the next chapter, we will explain some of the existing best practices used among Kubernetes projects and provoke forward-looking thoughts about how to prepare for the ongoing maintenance and development of your new Operator.

8
Preparing for Ongoing Maintenance of Your Operator

In this book, we have shown the steps for creating a new **Kubernetes** Operator. We've covered the full range of processes from conception to design, to coding, deployment, and finally, release. But, very few software projects end their life cycle after the initial release, Operators included. In fact, for many Operators, the majority of work will eventually be done after release on a long enough timescale. So, it's valuable to prepare for the future maintenance of your Operator by understanding the expectations of your users and the Operator community at large.

As a Kubernetes-based project, it can be very helpful to rely on the established conventions from Kubernetes and its subprojects for your own ongoing development. While you are free to form your own guidelines for future releases, it is likely that your Operator will depend on at least some library or aspect of the core Kubernetes platform simply by virtue of being built for Kubernetes. For that reason, knowing the policies in place for Kubernetes can help align your own development practices in preparation for changes in the upstream platform, which you will almost certainly inherit and be forced to react to. That's why this chapter will focus on those conventions throughout the following sections:

- Releasing new versions of your Operator
- Planning for deprecation and backward compatibility
- Complying with Kubernetes standards for changes
- Aligning with the Kubernetes release timeline
- Working with the Kubernetes community

The procedures established by the Kubernetes community provide an excellent template for your own development practices and a familiar set of policies for your users. There is, of course, no requirement for any Operator to strictly follow these guidelines, but it is the goal of this chapter to explain them in a way that provides the relevant precedent for your project.

Technical requirements

For this chapter, the only technical work will be done in the *Releasing new versions of your Operator* section, in which we will build on the existing nginx Operator code from earlier chapters to add new code and run that code in a Kubernetes cluster. As such, the requirements for this chapter include the following:

- The `operator-sdk` binary
- Go 1.16+
- Docker
- Access to a running Kubernetes cluster

The Code in Action video for this chapter can be viewed at: `https://bit.ly/3aiaokl`

Releasing new versions of your Operator

Now that your Operator is published, the fun has just begun. It's now time to start thinking about your next release! As with any software project, your Operator will evolve over time by introducing new features and adapting to changes in upstream Kubernetes. There are tomes of literature written on releasing software with recommendations on when and how to publish updates to a software product. Much of that information is out of the scope of this book. Instead, we will explain the technical steps required to create and publish a new version of your Operator with the **Operator SDK**, **Operator Lifecycle Manager** (**OLM**), and **OperatorHub** in mind. From there, methods and timing for your release are entirely up to your organization (though you may want to learn more about how other Kubernetes projects are released later in this chapter in the *Aligning with the Kubernetes release timeline* section).

Adding an API version to your Operator

While there are many factors that may influence your decision to release a new version of your Operator (such as bug fixes or simply following a regular release schedule), one type of change that is common among Operators is updating the Operator's config API. Recall that this is the API translated into the Operator's **CustomResourceDefinition** (**CRD**). So, in some cases, it may be necessary to update the API that is shipped with your Operator to indicate changes to users (see the *Complying with Kubernetes standards for changes* section).

To do this, we need to create a new API version and include that version in the Operator's CRD (for a more in-depth look at what this means from a technical standpoint, see the Kubernetes documentation on CRD versioning for information on the details of how this works: https://kubernetes.io/docs/tasks/extend-kubernetes/custom-resources/custom-resource-definition-versioning/).

In *Chapter 4*, *Developing an Operator with the Operator SDK*, we initialized the first version of our Operator's API with the following command:

```
operator-sdk create api --group operator --version v1alpha1
--kind NginxOperator --resource --controller
```

This created the NginxOperator API type as the v1alpha1 version. We then filled out the API in api/v1alpha1/nginxoperator_types.go and generated the corresponding CRD, which provides the interface for using the Operator once it was deployed in a cluster.

If some incompatible changes needed to be made to this API that required a new version, that version could be generated similarly. Say, for example, that we wanted to allow the nginx Deployment managed by our Operator to expose multiple ports, such as one for HTTP and another for HTTPS requests. We could do this by changing the port field in the existing nginx Operator's CRD to a ports field that defines a list of v1.ContainerPorts (this is a native Kubernetes API type that allows for naming multiple ports for a container). This new type exposes additional information, such as Name and HostPort, but it also includes the same int32 value for defining a singular ContainerPort that the original port field used. Let's take the following line from controllers/nginxoperator_controller.go as an example:

```
func (r *NginxOperatorReconciler) Reconcile(ctx context.
Context, req ctrl.Request) (ctrl.Result, error) {
   if operatorCR.Spec.Port != nil {
      deployment.Spec.Template.Spec.Containers[0].Ports[0].
ContainerPort = *operatorCR.Spec.Port
   }
```

This could be simplified to just the following:

```
func (r *NginxOperatorReconciler) Reconcile(ctx context.
Context, req ctrl.Request) (ctrl.Result, error) {
   if len(operatorCR.Spec.Ports) > 0 {
      deployment.Spec.Template.Spec.Containers[0].Ports =
operatorCR.Spec.Ports
   }
```

To show what this means for the Operator types, we will take the existing NginxOperatorSpec type in v1alpha1:

```
// NginxOperatorSpec defines the desired state of NginxOperator
type NginxOperatorSpec struct {

   // Port is the port number to expose on the Nginx Pod
   Port *int32 `json:"port,omitempty"`

   // Replicas is the number of deployment replicas to scale
   Replicas *int32 `json:"replicas,omitempty"`
```

```
    // ForceRedploy is any string, modifying this field
instructs
    // the Operator to redeploy the Operand
    ForceRedploy string `json:"forceRedploy,omitempty"`
}
```

And now, we will change it to a new NginxOperatorSpec type in v1alpha2 that looks like the following:

```
// NginxOperatorSpec defines the desired state of NginxOperator
type NginxOperatorSpec struct {

    // Ports defines the ContainerPorts exposed on the Nginx Pod
    Ports []v1.ContainerPort `json:"ports,omitempty""`

    // Replicas is the number of deployment replicas to scale
    Replicas *int32 `json:"replicas,omitempty"`

    // ForceRedploy is any string, modifying this field
instructs
    // the Operator to redeploy the Operand
    ForceRedploy string `json:"forceRedploy,omitempty"`
}
```

In order to preserve functionality for our users, it's important to introduce the new version in a way that ensures the Operator supports both versions for as long as required by the deprecation policy.

Generating the new API directory

The first step is to generate the new API files. The new API version is generated as a scaffold with the operator-sdk command, just like when we generated v1alpha1:

```
$ operator-sdk create api --group operator --version v1alpha2
--kind NginxOperator --resource
Create Controller [y/n]
n
Writing kustomize manifests for you to edit...
Writing scaffold for you to edit...
```

```
api/v1alpha2/nginxoperator_types.go
Update dependencies:
$ go mod tidy
Running make:
$ make generate
/Users/mdame/nginx-operator/bin/controller-
genobject:headerFile="hack/boilerplate.go.txt" paths="./..."
Next: implement your new API and generate the manifests (e.g.
CRDs,CRs) with:
$ make manifests
```

Note that this time, the `--controller` flag was omitted (and we chose n for `Create Controller [y/n]`) because the controller for this Operator already exists (`controllers/nginxoperator_controller.go`), so we don't need to generate another one.

Instead, the existing controller will need to be manually updated to remove references to `v1alpha1` and replace them with `v1alpha2`. This step can also be automated with tools such as `sed`, but be sure to carefully review any changes whenever automating code updates like this.

When the version is generated, it will create a new `api/v1alpha2` folder, which also contains an `nginxoperator_types.go` file. Copy the existing type definitions from `api/v1alpha1/nginxoperator_types.go` into this file and change the `port` field into `ports`, as shown in the preceding code. The new file should look as follows (note the highlighted change to `Ports`):

api/v1alpha2/nginxoperator_types.go:

```
package v1alpha2

import (
    v1 "k8s.io/api/core/v1"
    metav1 "k8s.io/apimachinery/pkg/apis/meta/v1"
)

const (
    ReasonCRNotAvailable           =
"OperatorResourceNotAvailable"
```

```go
	ReasonDeploymentNotAvailable  =
"OperandDeploymentNotAvailable"
	ReasonOperandDeploymentFailed = "OperandDeploymentFailed"
	ReasonSucceeded               = "OperatorSucceeded"
)


// NginxOperatorSpec defines the desired state of NginxOperator
type NginxOperatorSpec struct {
	// INSERT ADDITIONAL SPEC FIELDS - desired state of cluster
	// Important: Run "make" to regenerate code after modifying
this file

	// Ports defines the ContainerPorts exposed on the Nginx Pod
	Ports []v1.ContainerPort `json:"ports,omitempty""`

	// Replicas is the number of deployment replicas to scale
	Replicas *int32 `json:"replicas,omitempty"`

	// ForceRedploy is any string, modifying this field
instructs
	// the Operator to redeploy the Operand
	ForceRedploy string `json:"forceRedploy,omitempty"`
}


// NginxOperatorStatus defines the observed state of
NginxOperator
type NginxOperatorStatus struct {
	// Conditions is the list of the most recent status
condition updates
	Conditions []metav1.Condition `json:"conditions"`
}


//+kubebuilder:object:root=true
//+kubebuilder:subresource:status
//+kubebuilder:storageversion


// NginxOperator is the Schema for the nginxoperators API
```

```
type NginxOperator struct {
    metav1.TypeMeta    `json:",inline"`
    metav1.ObjectMeta `json:"metadata,omitempty"`

    Spec   NginxOperatorSpec    `json:"spec,omitempty"`
    Status NginxOperatorStatus `json:"status,omitempty"`
}


//+kubebuilder:object:root=true


// NginxOperatorList contains a list of NginxOperator
type NginxOperatorList struct {
    metav1.TypeMeta `json:",inline"`
    metav1.ListMeta `json:"metadata,omitempty"`
    Items              []NginxOperator `json:"items"`
}


func init() {
    SchemeBuilder.Register(&NginxOperator{},
&NginxOperatorList{})
}
```

Updating the Operator's CRD

Next, the Operator's CRD needs to be updated to include definitions for both
v1alpha1 and v1alpha2. First, one version needs to be defined as the **storage
version**. This is the version representation in which the object is persisted in etcd.
When the Operator only had one version, it wasn't necessary to specify this (that
was the only version available to be stored). Now, however, the API server needs to
know how to store the object. This is done by adding another kubebuilder marker
(//+kubebuilder:storageversion) to the NginxOperator struct:

```
//+kubebuilder:object:root=true
//+kubebuilder:subresource:status
//+kubebuilder:storageversion


// NginxOperator is the Schema for the nginxoperators API
type NginxOperator struct {
```

```
    metav1.TypeMeta    `json:",inline"`
    metav1.ObjectMeta `json:"metadata,omitempty"`

    Spec    NginxOperatorSpec    `json:"spec,omitempty"`
    Status NginxOperatorStatus `json:"status,omitempty"`
}
```

This instructs the CRD generator to label v1alpha2 as the storage version. Now, running make manifests will generate new changes to the CRD:

```
$ make manifests
$ git status
On branch master
Changes not staged for commit:
    modified:    config/crd/bases/operator.example.com_
nginxoperators.yaml
```

Now, the Operator's CRD should include a new v1alpha2 specification definition under versions:

config/crd/bases/operator.example.com_nginxoperators.yaml:

```
apiVersion: apiextensions.k8s.io/v1
kind: CustomResourceDefinition
metadata:
  annotations:
    controller-gen.kubebuilder.io/version: v0.7.0
  creationTimestamp: null
  name: nginxoperators.operator.example.com
spec:
  group: operator.example.com
  names:
    kind: NginxOperator
    listKind: NginxOperatorList
    plural: nginxoperators
    singular: nginxoperator
  scope: Namespaced
  versions:
```

```
  - name: v1alpha1
    schema:
      openAPIV3Schema:

        . . .

    served: true
    storage: false
    subresources:
      status: {}
  - name: v1alpha2
    schema:
      openAPIV3Schema:

        . . .

    served: true
    storage: true
    subresources:
      status: {}
```

Implementing API conversions

Finally, the API server needs to know how to convert between these two incompatible versions. Specifically, the `int32` value for the `v1alpha1` port needs to be converted into a `ContainerPort` value in a list of `ports` for `v1alpha2`. For the sake of this example, we will define the behavior as the following:

- `v1alpha1` to `v1alpha2`: `int32(port)` becomes `ports[0].ContainerPort`.

- `v1alpha2` to `v1alpha1`: `ports[0].ContainerPort` becomes `int32(port)`.

In other words, if we are handed a list of ports and need to convert to a single port (`v1alpha2` to `v1alpha1`), we will take the first value in the list and use that. In reverse (`v1alpha1` to `v1alpha2`), we will take the single `port` value and make it the first (and only) value in a new list of ports.

To define these conversion rules, we will implement the `Convertible` and `Hub` interfaces from `sigs.k8s.io/controller-runtime/pkg/conversion`:

---

sigs.k8s.io/controller-runtime/pkg/conversion/conversion.go:

```
package conversion
import "k8s.io/apimachinery/pkg/runtime"
// Convertible defines capability of a type to convertible i.e.
it can be converted to/from a hub type.
type Convertible interface {
      runtime.Object
      ConvertTo(dst Hub) error
      ConvertFrom(src Hub) error
}
// Hub marks that a given type is the hub type for conversion.
This means that
// all conversions will first convert to the hub type, then
convert from the hub
// type to the destination type. All types besides the hub type
should implement
// Convertible.
type Hub interface {
      runtime.Object
      Hub()
}
```

These will then be exposed to the API server via a **webhook** that creates an endpoint that can be queried by the API server to instruct it on how to perform conversions between different API versions of an object. The steps for creating a webhook with `operator-sdk` are wrappers for kubebuilder commands, so the steps for implementing the conversion webhook in an Operator are the same as for any other controller, as shown in the kubebuilder documentation). This process defines one version as the `Hub` type through which the `Convertible` spoke types are converted.

To get started, first create a new file, api/v1alpha2/nginxoperator_ conversion.go, to define v1alpha2 as the Hub version:

api/v1alpha2/nginxoperator_conversion.go:

```
package v1alpha2


// Hub defines v1alpha2 as the hub version
func (*NginxOperator) Hub() {}
```

Next, create another file, api/v1alpha1/nginxoperator_conversion.go (note this is in the v1alpha1 directory). This file will implement the ConvertTo() and ConvertFrom() functions to translate v1alpha1 to and from v1alpha2:

api/v1alpha1/nginxoperator_conversion.go:

```
package v1alpha1

import (
    "github.com/sample/nginx-operator/api/v1alpha2"
    v1 "k8s.io/api/core/v1"
    "k8s.io/utils/pointer"
    "sigs.k8s.io/controller-runtime/pkg/conversion"
)


// ConvertTo converts v1alpha1 to v1alpha2
func (src *NginxOperator) ConvertTo(dst conversion.Hub) error {
    return nil
}
// ConvertFrom converts v1alpha2 to v1alpha1
func (dst *NginxOperator) ConvertFrom(src conversion.Hub) error
{
    return nil
}
```

Then, fill in these functions to do the actual conversions. For the fields such as `Replicas` and `ForceRedeploy`, the conversion is 1:1 (it's also important to copy the `Metadata` and `Status.Conditions` too). But, for `Port/Ports`, we need to add the logic defined previously. That makes `ConvertTo()` look like the following:

```go
// ConvertTo converts v1alpha1 to v1alpha2
func (src *NginxOperator) ConvertTo(dst conversion.Hub) error {
    objV1alpha2 := dst.(*v1alpha2.NginxOperator)
    objV1alpha2.ObjectMeta = src.ObjectMeta
    objV1alpha2.Status.Conditions = src.Status.Conditions

    if src.Spec.Replicas != nil {
        objV1alpha2.Spec.Replicas = src.Spec.Replicas
    }
    if len(src.Spec.ForceRedploy) > 0 {
        objV1alpha2.Spec.ForceRedploy = src.Spec.ForceRedploy
    }
    if src.Spec.Port != nil {
        objV1alpha2.Spec.Ports = make([]v1.ContainerPort, 0, 1)
        objV1alpha2.Spec.Ports = append(objV1alpha2.Spec.Ports,
            v1.ContainerPort{ContainerPort: *src.Spec.Port})
    }

    return nil
}
```

And `ConvertFrom()` is similar, but in reverse:

```go
// ConvertFrom converts v1alpha2 to v1alpha1
func (dst *NginxOperator) ConvertFrom(src conversion.Hub) error
{
    objV1alpha2 := src.(*v1alpha2.NginxOperator)
    dst.ObjectMeta = objV1alpha2.ObjectMeta
    dst.Status.Conditions = objV1alpha2.Status.Conditions

    if objV1alpha2.Spec.Replicas != nil {
        dst.Spec.Replicas = objV1alpha2.Spec.Replicas
    }
```

```
    if len(objV1alpha2.Spec.ForceRedploy) > 0 {
        dst.Spec.ForceRedploy = objV1alpha2.Spec.ForceRedploy
    }


    if len(objV1alpha2.Spec.Ports) > 0 {
        dst.Spec.Port = pointer.Int32(objV1alpha2.Spec.Ports[0].
ContainerPort)
    }


    return nil
}
```

Now, we can generate the webhook logic and endpoints by using `operator-sdk create webhook`, with the following command:

```
$ operator-sdk create webhook --conversion --version v1alpha2
--kind NginxOperator --group operator --force
Writing kustomize manifests for you to edit...
Writing scaffold for you to edit...
api/v1alpha2/nginxoperator_webhook.go
Webhook server has been set up for you.
You need to implement the conversion.Hub and conversion.
Convertible interfaces for your CRD types.
Update dependencies:
$ go mod tidy
Running make:
$ make generate
/Users/mdame/nginx-operator/bin/controller-gen
object:headerFile="hack/boilerplate.go.txt" paths="./..."
Next: implement your new Webhook and generate the manifests
with:
$ make manifests
```

You can ignore the message saying `You need to implement the conversion. Hub and conversion.Convertible interfaces for your CRD types` because we already did that (the generator simply assumes that it will be run before these are implemented). At this point, everything has been implemented, and the only next step is to ensure that the webhook is properly deployed with the Operator when it is installed in a cluster.

Updating project manifests to deploy the webhook

Just like how the manifests to enable metrics resources needed to be uncommented in order to be deployed with the Operator (*Chapter 5*, *Developing an Operator – Advanced Functionality*), so too do resources related to the webhook.

To do this, first, modify `config/crd/kustomization.yaml` to uncomment the `patches/webhook_in_nginxoperators.yaml` and `patches/cainject_in_nginxoperators.yaml` lines to include these two patch files in the deployment:

config/crd/kustomization.yaml:

```
resources:
- bases/operator.example.com_nginxoperators.yaml
#+kubebuilder:scaffold:crdkustomizeresource

patchesStrategicMerge:
# [WEBHOOK] To enable webhook, uncomment all the sections with
[WEBHOOK] prefix.
# patches here are for enabling the conversion webhook for each
CRD
- patches/webhook_in_nginxoperators.yaml
#+kubebuilder:scaffold:crdkustomizewebhookpatch

# [CERTMANAGER] To enable cert-manager, uncomment
all the sections with [CERTMANAGER] prefix.
# patches here are for enabling the CA injection for each CRD
- patches/cainjection_in_nginxoperators.yaml
#+kubebuilder:scaffold:crdkustomizecainjectionpatch

# the following config is for teaching kustomize how to do
kustomization for CRDs.
configurations:
- kustomizeconfig.yaml
```

Now, modify one of those files, `patches/webhook_in_nginxoperators.yaml`, to add the two CRD versions as `conversionReviewVersions` in the Operator's CRD:

config/crd/patches/webhook_in_nginxoperators.yaml:

```
# The following patch enables a conversion webhook for the CRD
apiVersion: apiextensions.k8s.io/v1
kind: CustomResourceDefinition
metadata:
  name: nginxoperators.operator.example.com
spec:
  conversion:
    strategy: Webhook
    webhook:
      clientConfig:
        service:
          namespace: system
          name: webhook-service
          path: /convert
      conversionReviewVersions:
        - v1
        - v1alpha1
        - v1alpha2
```

Next, make the following changes to `config/default/kustomization.yaml` to uncomment the following lines:

- `- ../webhook`
- `- ../certmanager`
- `- manager_webhook_patch.yaml`
- All of the variables in the `vars` section with the `[CERTMANAGER]` label.

The final file will look as follows (uncommented lines highlighted and some sections omitted for brevity):

---

config/default/kustomization.yaml:

```
...
bases:
- ../crd
- ../rbac
- ../manager
# [WEBHOOK] To enable webhook, uncomment all the sections with
[WEBHOOK] prefix including the one in
# crd/kustomization.yaml
- ../webhook
# [CERTMANAGER] To enable cert-manager, uncomment all sections
with 'CERTMANAGER'. 'WEBHOOK' components are required.
- ../certmanager
# [PROMETHEUS] To enable prometheus monitor, uncomment all
sections with 'PROMETHEUS'.
- ../prometheus
...
# [WEBHOOK] To enable webhook, uncomment all the sections with
[WEBHOOK] prefix including the one in
# crd/kustomization.yaml
- manager_webhook_patch.yaml
# [CERTMANAGER] To enable cert-manager, uncomment all sections
with 'CERTMANAGER'.
# Uncomment 'CERTMANAGER' sections in crd/kustomization.yaml to
enable the CA injection in the admission webhooks.
# 'CERTMANAGER' needs to be enabled to use ca injection
#- webhookcainjection_patch.yaml
# the following config is for teaching kustomize how to do var
substitution
vars:
# [CERTMANAGER] To enable cert-manager, uncomment all sections
with 'CERTMANAGER' prefix.
- name: CERTIFICATE_NAMESPACE # namespace of the certificate CR
  objref:
```

```
    kind: Certificate
    group: cert-manager.io
    version: v1
    name: serving-cert # this name should match the one in
certificate.yaml
  fieldref:
    fieldpath: metadata.namespace
- name: CERTIFICATE_NAME
  objref:
    kind: Certificate
    group: cert-manager.io
    version: v1
    name: serving-cert # this name should match the one in
certificate.yaml
- name: SERVICE_NAMESPACE # namespace of the service
  objref:
    kind: Service
    version: v1
    name: webhook-service
  fieldref:
    fieldpath: metadata.namespace
- name: SERVICE_NAME
  objref:
    kind: Service
    version: v1
    name: webhook-service
```

And finally, comment out the `manifests.yaml` line in `config/webhook/`
`kustomization.yaml` (this file does not exist for our use case, and trying to deploy
without uncommenting this line will result in an error). The following snippet shows
which line should be commented out with #:

---

config/webhook/kustomization.yaml:

```
  resources:
#- manifests.yaml

  - service.yaml
```

With these changes, the Operator can be re-built and deployed using the `operator-sdk` and `make` commands from earlier chapters.

Deploying and testing the new API version

To confirm that the API server now understands and can convert between versions of the Operator CRD, install it in a cluster. Note that now your Operator will depend on **cert-manager** being present in the cluster, so be sure to install that first (instructions are available at `https://cert-manager.io/docs/installation/`).

Remember, you need to update `controllers/nginxoperator_controller.go` to replace `v1alpha1` references with `v1alpha2` and change the `Ports` check (in `Reconcile()`) to match the following:

controllers/nginxoperator_controller.go:

```
func (*NginxOperatorReconciler) Reconcile(…) {
  if len(operatorCR.Spec.Ports) > 0 {
   deployment.Spec.Template.Spec.Containers[0].Ports =
operatorCR.Spec.Ports
  }
}
```

Forgetting to do this will cause an error when creating or retrieving the Operator CRD (which won't show up at compile time). This is because the `v1alpha1` API types are still defined and valid (so the Operator code compiles correctly), but the new client and reconciliation code will expect the object to be retrieved in the `v1alpha2` format.

To deploy the nginx Operator, build and push the new container image before running `make deploy`:

```
$ export IMG=docker.io/mdame/nginx-operator:v0.0.2
$ make docker-build docker-push
$ make deploy
```

Next, create a simple `NginxOperator` object. To demonstrate the API conversion, create it as the `v1alpha1` version and use the old `port` field:

sample-cr.yaml:

```yaml
apiVersion: operator.example.com/v1alpha1
kind: NginxOperator
metadata:
  name: cluster
  namespace: nginx-operator-system
spec:
  replicas: 1
  port: 8080
```

Next, create the custom resource object with kubectl:

```
$ kubectl apply -f sample-cr.yaml
```

Now, using `kubectl get` to view the object will show it as `v1alpha2` because it was automatically converted and stored as this version:

```yaml
$ kubectl get -o yaml nginxoperators/cluster -n nginx-operator-
system
apiVersion: operator.example.com/v1alpha2
kind: NginxOperator
metadata:
  ...
  name: cluster
  namespace: nginx-operator-system
  resourceVersion: "9032"
  uid: c22f6e2f-58a5-4b27-be6e-90fd231833e2
spec:
  ports:
  - containerPort: 8080
    protocol: TCP
  replicas: 1
...
```

You can choose to specifically view the object as `v1alpha1` with the following command, which will instruct the API server to call the Operator's webhook and convert it back using the functions we wrote:

```
$ kubectl get -o yaml nginxoperators.v1alpha1.operator.example.
com/cluster -n nginx-operator-system
apiVersion: operator.example.com/v1alpha1
kind: NginxOperator
metadata:
  name: cluster
  namespace: nginx-operator-system
  resourceVersion: "9032"
  uid: c22f6e2f-58a5-4b27-be6e-90fd231833e2
spec:
  port: 8080
  replicas: 1
```

What this means for your users is that they can continue using the existing API, which allows valuable transition time while you introduce a new version. Note that if they are already using the Operator, and you introduce a new storage version, they may need to use **kube-storage-version-migrator** (`https://github.com/kubernetes-sigs/kube-storage-version-migrator`) to migrate the existing storage version to the new one. You can provide migration files for them (or even automate it into the Operator, as migrations are simply Kubernetes resources) to make this easier.

With a new API version introduced and conversion handled, your Operator is now ready to be packaged into a new bundle so that deployment can be managed by the OLM. This means updating your Operator's CSV to a new version.

Updating the Operator CSV version

Updating the version of the Operator in its CSV provides information to the OLM, OperatorHub, and users about which version of the Operator they are running. It also instructs the OLM on which versions replace other versions for in-cluster upgrades. This allows developers to define specific upgrade **channels** (such as `alpha` and `beta`), which are similar to the versioning channels in other software projects that allow users to subscribe to a different release cadence. The Operator SDK documentation goes into technical detail about this process on GitHub, but it is not necessary to understand these details for completing this section (`https://github.com/operator-framework/operator-lifecycle-manager/blob/b43ecc4/doc/design/how-to-update-operators.md`). In this section, however, we will cover the simple task of updating the CSV version in a single channel.

The first step toward bumping the Operator's CSV version is updating the version that will be replaced with the current version. In other words, `v0.0.2` will replace `v0.0.1`, so the `v0.0.2` CSV must indicate that it is replacing `v0.0.1`.

This is done by modifying the base CSV in `config/manifests/bases` to add a `replaces` field under its spec, as shown in the following example:

config/manifests/bases/nginx-operator.clusterserviceversion.yaml:

```
apiVersion: operators.coreos.com/v1alpha1
kind: ClusterServiceVersion
metadata:
  annotations:
    alm-examples: '[]'
    capabilities: Basic Install
  name: nginx-operator.v0.0.0
  namespace: placeholder
spec:
  ...
  replaces: nginx-operator.v0.0.1
```

Next, update the VERSION variable in the project's Makefile (you can also export this variable to the new version in your shell similar to the other environment variables we have used from this file, but updating it manually clearly indicates the version and ensures that the right version will be propagated when built on any machine):

Makefile:

```
# VERSION defines the project version for the bundle.

# Update this value when you upgrade the version of your
project.
# To re-generate a bundle for another specific version without
changing the standard setup, you can:
# - use the VERSION as arg of the bundle target (e.g make
bundle VERSION=0.0.2)
# - use environment variables to overwrite this value (e.g
export VERSION=0.0.2)
VERSION ?= 0.0.2
```

Now, the new CSV can be built as part of the regular make bundle command from *Chapter 7, Installing and Running Operators with the Operator Lifecycle Manager*:

```
$ make bundle IMG=docker.io/sample/nginx-operator:v0.0.2
```

This updates the version of the Operator listed in bundle/manifests/nginx-operator.clusterserviceversion.yaml (which is the main CSV packaged into the Operator's bundle). If you followed the steps from the previous section to add a new API version, it also adds information about that new version (and the conversion webhook) to the CSV and the sample CRD packaged along with the bundle. In addition, it will generate a new Service manifest in the bundle that exposes the conversion endpoint for the two API versions.

The bundle image can then be built, pushed, and run with the OLM just like before:

```
$ export BUNDLE_IMG=docker.io/sample/nginx-bundle:v0.0.2
$ make bundle-build bundle-push
$ operator-sdk run bundle docker.io/same/nginx-bundle:v0.0.2
```

With a new bundle image built and working, the only step remaining in releasing your new version is to publish it on OperatorHub for users to find.

Releasing a new version on OperatorHub

With a new version of your Operator, the bundle that is published to OperatorHub also needs to be updated (if you have chosen to release your Operator on OperatorHub). Thankfully, this process is not too complex. In fact, it is essentially the same as releasing your initial version (as in *Chapter 7, Installing and Running Operators with the Operator Lifecycle Manager*), wherein you created a folder with your Operator's bundle and submitted that folder as a pull request to the community Operators repository on GitHub (`https://github.com/k8s-operatorhub/community-operators`).

To release a new version of your Operator, simply create a new folder under your Operator's directory in the community Operators project named after the version number. For example, if the first version was `nginx-operator/0.0.1`, this version would be `nginx-operator/0.0.2`.

Then, just like with your first version, simply copy the contents of your Operator's `bundle` directory (after generating the new bundle version) into the new version folder. Commit and push the changes to your fork of the GitHub repository and open a new pull request against the main repository with the changes.

When your pull request passes the automated checks, it should merge, and the new version of your Operator should be visible on OperatorHub soon after.

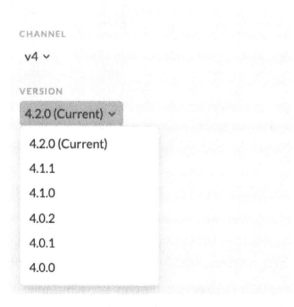

Figure 8.1 – Screenshot of version and channel listing on OperatorHub for the Grafana Operator

Now, you have finished releasing a new version of your Operator. By introducing a new API, ensuring that the new API is convertible between existing versions, updating your Operator's bundle, and publishing that updated bundle on OperatorHub, you should now announce to your users that the new version of your Operator is available. In the next sections, we'll discuss ways to ensure that future releases continue to go smoothly by planning ahead to minimize breaking API changes that require new versions and following the Kubernetes standards for making new changes.

Planning for deprecation and backward compatibility

In the previous section, we discussed the work needed to release a new version of an Operator. While the processes for bundling and publishing a new version are relatively simple in terms of the effort required, implementing a new API version is not an insignificant task. As such, doing so should be done only as necessary in order to minimize the use of engineering resources and disruption to users.

Of course, it will occasionally be unavoidable that incompatible changes must be introduced, for example, in the case of deprecation. Some of this deprecation might even come from upstream, where it is beyond your direct control (see the *Complying with Kubernetes standards for changes* section). However, the frequency of such changes can often be controlled through careful planning. In this section, we'll discuss some ways to plan for deprecation and support backward compatibility without causing undue strain on your engineers or users.

Revisiting Operator design

In *Chapter 2, Understanding How Operators Interact with Kubernetes*, the *Planning for changes in your Operator* section discussed, in general terms, various design approaches that establish good practices for outlining an Operator's design for future evolution. These suggested guidelines (which could, in reality, be applied to many software projects) were to start small, iterate effectively, and deprecate gracefully.

Now, having gone through the steps of building an Operator from scratch in the form of our nginx Operator, it will be helpful to revisit these guidelines and examine how they were specifically applied to our own design.

Starting small

The nginx Operator design started out very simple. The Operator was intended to serve a very basic function: manage the deployment of an nginx Pod. For configuration options, it exposed three fields to control the container port on the Pod, the number of replicas to deploy, and an extra field to trigger a forced redeployment. As a **minimum viable product** (**MVP**), this served well to get our Operator off the ground. While this design was intentionally kept minimal for the purpose of a reasonably sized demonstration, it still shows the mindset that keeps Operator CRDs from starting out with too many knobs exposed. Shipping excessive configuration options can be confusing to users, which then poses potential risks to the stability of your product if users cannot fully understand what every option does. This also adds a support burden on your own teams.

Remember that the first release of an Operator will actually likely make up a minority of its life cycle. There will be plenty of time to add more features in future releases, but this becomes difficult as the CRD grows in its API definition and new features dilute the viability of existing ones. Of course, that's not to say you should never add new features to your Operator or its CRD. But, when that time comes, it's important to do so carefully, which is the focus of iterating effectively.

Iterating effectively

In this chapter, we introduced an API change to the Operator wherein the type of one field was converted to an entirely different nested struct. This effectively removed the old field, which would be a breaking change for any user who was already relying on that field (in fact, we did technically remove the old field, but more on that in a minute). The benefit of changes such as this needs to be weighed against the negatives, which include disruption for your users (if applicable) and ongoing support for your own team.

In our case, we chose to move from a single `int32` field (named `port`) to a list of `v1.ContainerPort` objects (named `ports`). This added some complexity to the CRD (because `v1.ContainerPort` contains other nested fields). However, it also shifted our CRD to rely on a native, stable API type from upstream Kubernetes. This, along with the added functionality of being able to configure multiple ports and added fields, offers usability and stability benefits for users and developers of the Operator (not that `int32` is likely to be unstable, but the general idea remains).

Still, this change required the removal of the existing `port` field. This is what users must ultimately react to, but that transition can be made smoother through graceful deprecation.

Deprecating gracefully

When adding the new `ports` field to our Operator's CRD, a conscious decision was made to remove the existing single `port` field, and that's perfectly acceptable; in fact, keeping it would have been unnecessarily redundant. But, the fact remains that users who were relying on the old field would have to transition to the new one or face data loss when their old configurations were migrated to the new ones. While that may seem small for something such as a single integer value, the scope of degradation can clearly scale with more complex Operators.

This is why we added a conversion webhook to the Operator, to translate the old field into the new one automatically. But that webhook would not have been as simple to implement if the new `ports` field wasn't effectively a super-set of the old field. Choosing to go with a fairly compatible existing type made this transition much easier for developers to implement and users to understand. Design decisions such as this greatly help to reduce friction in a growing code base.

However, our conversion was not completely perfect. While the transition from `v1alpha1` to `v1alpha2` carries over just fine, the reverse can only hold one port value (the first from the list). This may be suitable for practical use cases, as most users would be more likely to upgrade to the new version than downgrade to the deprecated one, but from a support standpoint, lossy conversion like this can create headaches down the road. There are ways to address this that are relevant to the next section, which discusses how Kubernetes prescribes ways to smoothly implement changes.

Complying with Kubernetes standards for changes

Kubernetes defines a standard set of policies for deprecation (and other breaking changes) that all core projects must abide by. This policy is available at `https://kubernetes.io/docs/reference/using-api/deprecation-policy/`. It's not necessary to read and understand the whole policy for the purposes of Operator development, as we will highlight some of the relevant bits here. It primarily defines the standards for deprecating parts of the Kubernetes API, with many of the same (or similar) guidelines being applied to other types of deprecation as well (such as user-facing features that are not directly part of the API). It does this by enumerating a list of explicit rules for deprecating changes, some of which we will cover in this section.

As a third-party component, your Operator is under no obligation to follow the Kubernetes deprecation policy. But, in a practical sense, there are benefits to working within the constraints of the ecosystem your Operator is built to be a part of. These include a template provided to set expectations for your users and a set of guidelines for planning your own ongoing development. And, even if you choose not to follow these policies for your own Operator, it is still crucial to understand how they are enforced upstream to be prepared for deprecations and changes beyond your control that you will eventually inherit.

The full deprecation policy linked at the start of this section goes into specific detail for standards that govern every applicable Kubernetes component. So, some details of the deprecation are not directly relevant to Operator development. However, certain elements, such as those pertaining to the support and removal of API fields, do apply to Operators (should you choose to follow them).

Removing APIs

The Kubernetes deprecation policy is very clear on forbidding the removal of API elements from the current API version. In fact, this is the first rule in the entire policy.

> **Rule #1**
> API elements may only be removed by incrementing the version of the API group.

This means that it is forbidden to remove any field or object from an existing API. The removal can only be done by introducing a new API version with that element removed.

This is relevant to our nginx Operator, where we removed the `port` field that was present in `v1alpha1` as we incremented the API version to `v1alpha2`. Following this rule ensures that current users of an API version will not have their workflows suddenly broken by updating to a new release. The distinction between API versions makes a clear indication of some level of incompatibility.

Inversely, this rule allows for the addition of API elements without incrementing the version of the existing API. This is because a new element in the current API version will not break any existing use cases because it is as if consumers of the object are simply leaving this field blank (as opposed to the removal of an existing field, which could result in data loss as non-blank entries are dropped). This is directly relevant to our use case because it provides an ability for seamless conversion.

API conversion

The second rule of the Kubernetes deprecation policy is as follows.

> **Rule #2**
>
> API objects must be able to round-trip between API versions in a given release without information loss, with the exception of whole REST resources that do not exist in some versions.

This means that when any two API versions exist within the same release of Kubernetes (or, in this case, your Operator), objects of either API version must be able to convert between the two versions while preserving all data fields.

In our nginx Operator, we did not follow this rule (because a `v1alpha2` CRD object with multiple ports defined cannot translate them all to a single `port` value). This is OK because, as a third-party project, we are not beholden to the upstream Kubernetes policies. But, from a practical standpoint, it would be useful for us and our users to support such lossless conversions. This could be done by adding the `ports` field to both `v1alpha2` and `v1alpha1`. Then, our conversion could have stored the additional ports in the new field when converting to `v1alpha1`. Existing code that only knows about the `v1alpha1` single `port` field may not be able to use these additional ports, but the important part is that the data is preserved during conversion. Alternatively, we could have simply stored the additional `ports` values as an annotation in the CRD object's metadata and read from this during conversion.

API lifetime

How long you choose to support an API version is entirely up to your organization's agreement with your users. The Kubernetes standards for support timelines vary based on the stability level of that API. The three levels of stability are alpha, beta, and general availability (GA).

In our nginx Operator, the API is currently in alpha. Technically, this means that there is no support guarantee for any duration. However, when working toward graduating an API to a more stable level, it is good practice to operate as if that API is already at the next level of stability. For beta APIs, this timeline is the longer of 9 months or three releases (see the *Aligning with the Kubernetes release timeline* section). APIs that are graduated to GA cannot be removed, but they can be marked as deprecated. The intent is that an API that is GA can be assumed stable for the lifetime of that major version of Kubernetes.

Removal, conversion, and lifetime are the three main points of the Kubernetes deprecation policy relevant to our Operator development. There are more details in the link provided at the top of this section. You can also follow the planned timeline for upcoming upstream API deprecation, which is published at `https://kubernetes.io/docs/reference/using-api/deprecation-guide/`, for each release.

While this section mentioned the Kubernetes release as a unit of time, we did not define exactly how much time that is. In the next section, we will explain exactly how the Kubernetes release breaks down and how it relates to your Operator development.

Aligning with the Kubernetes release timeline

Each new release of Kubernetes is driven by the hard work and dedication of many people from different companies and time zones. This process for publishing a new version of Kubernetes is, therefore, the product of an orchestration effort among component areas and **special interest groups** (**SIGs**) to ensure timely, stable releases. While there are occasional roadblocks and delays, the release timeline is, for the most part, a well-defined collaborative effort that strives to provide predictable updates that the Kubernetes ecosystem of users and downstream products can depend on.

When developing an Operator, you or your organization will likely have similar organized release efforts. You also want to provide a dependable and timely schedule for shipping updates to your users. And, while your own release schedule may differ from the Kubernetes timeline, there are still benefits to a keen understanding of how the upstream release works. For example, it offers you the ability to plan around specific dates at which a beta API will be promoted to GA, or an entirely new feature will become available that you can leverage in your own product to pass on to your users. On the other hand, it also outlines the remaining time that deprecated APIs have before they are removed entirely. As a product vendor, you can rely on this timeline to guide your release planning.

For these reasons, aligning your Operator's release timeline for updates with that of Kubernetes is a valuable effort and it's why we will explain that timeline more in this section.

Overview of a Kubernetes release

The Kubernetes release cycle is approximately 15 weeks long. As of Kubernetes 1.22, this has defined a goal of three releases per calendar year. Of course, three 15-week releases do not account for an entire 52-week-long year. This is because the release cycle allows for several breaks in work, including holidays, end-of-year travel, and events or conferences such as **KubeCon**. This timeline was decided on by the SIG Release team with input from the community, and you can read more details about the decision in a blog post at `https://kubernetes.io/blog/2021/07/20/new-kubernetes-release-cadence/`.

During the course of a single release, there are several key dates that signal important progress updates in that release. These include an **Enhancements Freeze, Code Freeze**, deadlines for documentation updates and blog posts, and **release candidate (RC)** version releases. The exact timeline for each release is posted in the SIG Release GitHub repository at `https://github.com/kubernetes/sig-release`. As an example, the Kubernetes 1.24 release cycle looked like the following, with some of these key dates highlighted:

	Sunday	Monday	Tuesday	Wednesday	Thursday	Friday	Saturday
1		Start of Release					
2							
3							
4					Enhancements Freeze		
5							
6							
7							
8							
9							
10							
11		Call for Exceptions		Retro			
12			Code Freeze				
13			Test Freeze / rc.0 release				
14						rc.1 release	
15			GA Release				
16				Retro	Retro		

Figure 8.2 – Kubernetes 1.24 release cycle calendar

Each of these dates plays an important role in the progression of an individual release. In the following sections, we'll explain what each of them means in more detail and how they can relate to your own release cycle for your Operator. Much of this information is also documented in the SIG Release repository at `https://github.com/kubernetes/sig-release/blob/master/releases/release_phases.md`.

Start of release

Fairly self-explanatory, this is the official date that the release cycle for the next version of Kubernetes begins. This provides a reference date from which all other steps in the cycle can be derived. Note that this is not the same date as the previous release's publication because of the buffer between release cycles, along with post-release processes such as **Retrospective**. So, there may already be work-in-progress for the current release before this date (see the *GA release/Code Thaw* subsection). However, just as any race needs a starting line, every release needs a date to denote its official beginning.

Enhancements Freeze

New features in Kubernetes take many forms. While all changes to the platform are valuable regardless of size, certain undertakings involve increased effort and a broadened scope of work. Examples include significant user-facing changes or architectural designs involving collaboration between multiple components and SIGs. Such features require additional management to ensure their successful rollout. At this point, the feature may be considered as an **enhancement** or **Kubernetes Enhancement Proposal** (**KEP**).

We have already mentioned several individual KEPs in this book. For example, in *Chapter 5, Developing an Operator – Advanced Functionality,* we referred to **KEP-1623**, which defined the standard `Condition` type for components to report their status (including our nginx Operator). All KEPs such as this are tracked in the *Kubernetes Enhancements* GitHub repository at `https://github.com/kubernetes/enhancements`.

The full details of the KEP process and repository are beyond the scope of this book, but the essential knowledge is that KEPs signify wide-scale design changes that can impact users or consumers of APIs. As the developer of an Operator, which is itself a component that may consume one or more upstream APIs (such as the `Condition` field), it is then important to be aware of the status of upcoming changes of this scale, as they may directly affect you. Providing this information is the goal of the KEP process.

During a release cycle, the Enhancements Freeze stage signifies the point at which all proposed KEPs for that release must be either accepted and committed toward implementing their changes or delayed for a future release. This is the critical *go/no-go* date for moving forward with a KEP for that release. And, while many in-progress KEPs are able to proceed at this point, there may be a reasonable justification why certain changes may not be able to commit to the release by this date. In these cases, the developers of an enhancement may request an **exception** during the *Call for Exceptions* period, described next.

For Operator developers, the Enhancements Freeze deadline is an important date to keep in mind during your own development cycle because, while KEPs are often used to introduce new features to Kubernetes, part of their definition is to outline the removal of other features (such as any that are being replaced). Being aware of any features you depend on being officially slated for removal helps to gauge the urgency with which you will need to react to that removal. On the other hand, knowing that a KEP missed the deadline for Enhancements Freeze can provide assurance that any related features planned for removal will continue to be supported for at least another release unless that plan receives an exception.

Call for Exceptions

If, during the course of a release, a certain feature is not yet ready to commit to Enhancement Freeze (or *Code Freeze*), the contributors and participating SIGs who are working on that enhancement may request an exception to the freeze deadlines. If approved, this allows those contributors a reasonable extension to accommodate the additional time needed to prepare the enhancement for release.

Enhancements seeking an exception need to meet certain criteria to ensure that they do not risk impacting the stability of the platform or delaying the release. Therefore, release managers evaluate each exception request based on its scope, the estimated extension time requested, and the time at which the request was submitted within the release cycle.

Exception requests and approvals are generally communicated via the participating SIG's mailing lists, Slack channels, and the specific KEP issue discussion page on GitHub. This is why it's important to monitor these communication channels for features your Operator implements or depends on. Just because a new feature missed Enhancements Freeze, that doesn't mean it won't be implemented until an exception has been officially denied. The same goes for planned removals, in the event that an exception is granted. Knowing whether the upstream SIG related to your Operator's function is going to request any exceptions after Enhancements Freeze (or Code Freeze) is a good signal for what to commit to in your Operator's development cycle.

Code Freeze

Code Freeze is the point during the release cycle at which all code changes must be complete and merged into the Kubernetes code base (besides any features that have been granted an exception). This signifies that no more changes will be accepted into the release unless they are critical to the stability of the platform.

As the developer of a downstream Operator, this is relevant to your own project timeline because it is the point at which you can begin updating your libraries to the latest Kubernetes version with a reasonable expectation of stability. This can be done by updating your dependencies to the latest RC version of the upstream Kubernetes libraries.

The first RC for a new release is usually published soon after the Code Freeze date. This timing allows developers to update dependencies with additional *soak time* to catch any additional updates or reactions that need to be made in order to function with the new version of Kubernetes before it is finally released. It is very beneficial to take advantage of this timing to minimize the delay between publishing your updated Operator after the Kubernetes version release. Due to the size and fluctuating nature of Kubernetes, it's recommended to upgrade any upstream dependencies regularly. Failure to do so is likely to result in snowballing technical debt and eventual lack of compatibility with newer versions of the platform.

Test Freeze

While Code Freeze defines a strict deadline for enhancements to have their implementations completely merged, Test Freeze allows an additional buffer to expand test coverage. This provides an opportunity to improve test cases once a feature has merged before release. After this date, the only changes allowed to any tests are to fix or remove tests that are consistently failing. This date may not have a consistent impact on your own release cycle, but it is a good signal to be aware of when monitoring the progress of a key enhancement.

GA release/Code Thaw

At last, the moment everyone has been waiting for. If there are a few problems discovered during the release that cause a delay, this is the date that Kubernetes releases its newest version. For developers, this means that the code in Kubernetes (at `https://github.com/kubernetes/kubernetes`) is updated with a new Git tag, which allows the opportunity to easily reference the release point at a definitive point in code (for example, when updating your Operator's dependencies in its `go.mod` file). In addition, the client libraries that support Kubernetes and its subprojects are also updated, including the following:

- `https://github.com/kubernetes/api` – Core API types used by the Kubernetes platform (imported as `k8s.io/api`)

- `https://github.com/kubernetes/apimachinery` – The library used to implement the encoding and decoding of API types in code (`k8s.io/apimachinery`)

- `https://github.com/kubernetes/client-go` – The Go client used by Operators and other programs to interact with Kubernetes resources programmatically (`k8s.io/client-go`)

These are just a few of the additional dependencies that, on GA release day, will be updated with a new Git tag. These dependencies are actually part of the core Kubernetes repository (`k8s.io/kubernetes`), but they are synced to symbolic canonical locations by an automated bot for better dependency management. This can sometimes cause a slight delay between core Kubernetes releasing its tag and the libraries updating with their own (which are usually more important to developers).

> **Kubernetes as a Dependency**
>
> The core Kubernetes repository at `k8s.io/kubernetes` (or `https://github.com/kubernetes/kubernetes`) contains all of the code needed by the core platform components. So, it may be tempting to import code from here directly. However, due to its complexity, it is not recommended to import code from `k8s.io/kubernetes` directly into your project as it can cause dependency issues that are difficult to resolve with Go modules and bring excess transitive dependencies into your project. Instead, it is best to rely on the component libraries (such as those listed previously), which are meant to be imported into outside projects.

After the new release is officially published, the Kubernetes project enters **Code Thaw**, which, as its name implies, is the opposite of Code Freeze. This means that code development for the next release can begin merging into the main branch of Kubernetes. So, working with code from the `master` (or `main`) branch at any point after this really means you are interacting with the `N+1` version code (where `N` is the current version of Kubernetes).

Retrospective

When the dust has settled, the SIG Release team takes time to review the release during **Retrospective**. The intent of Retrospective is to meet and discuss any roadblocks or concerns that occurred during that release and propose solutions to address them in the future. In addition, any exceptional successes are identified and praised. It is meant as a blameless process through which the end goal is simply to reduce friction for future releases. While the specific details of the Retrospective may not be directly relevant to the development of your Operator (the focus is generally more on the release infrastructure and processes that drive the release rather than specific feature changes), it can be a very informative way to be aware of future changes to the release cycle.

All of the above dates form the most important signals in the Kubernetes release cycle. Being aware of them can help inform your team about the status of the current features you depend on, and that information can be passed on in the form of the commitments you make to your users. Staying up to date with the latest Kubernetes releases is critical to avoid tiresome technical debt from piling up, as the Kubernetes project is constantly evolving within the support constraints it offers.

In addition, knowing where you are in the current release cycle also provides the opportunity to contribute to Kubernetes at meaningful times. In the next section, we will discuss ways you can do this and how it can benefit you.

Working with the Kubernetes community

This chapter focused heavily on the standards, policies, and timelines of Kubernetes as they relate to Operator development. While it might seem that these are fixed, prescriptive decrees, the reality is that they are fluid frameworks for development created through an open and fair process. This process is organized by contributors from different companies all over the world, and it is always open to new voices.

As a developer for the Kubernetes platform, you have inherent stock in the community that organizes Kubernetes upstream. Therefore, improvements or concerns that affect you are likely to affect others as well. This is why it's not only OK but encouraged that you play an active role in upstream development alongside your own Operator. If nothing else, doing so serves to benefit your own development, as the open process allows you to help steer upstream work as you feel it should be done.

Getting involved is as simple as sending a message or joining a video call. The various GitHub repositories and Slack channels shown in other chapters for support are a great place to offer support yourself. The Kubernetes Slack server is `slack.k8s.io`, and it is free to join and contribute. You may also want to follow the various SIG meetings for topics of interest, all of which are listed on the *Kubernetes Community* repository at `https://github.com/kubernetes/community`. This repository includes links and schedules for all SIG and community meetings.

Summary

Once, in an interview, champion football quarterback Tom Brady was asked which of his championship rings was his favorite, to which he replied, "The next one." Coming from a person who many already considered to be one of the most successful in his field, that response showed powerful commitment to a continuous pursuit of achievement (if not a little tongue-in-cheek hubris). As software developers, that same passion is what drives the never-ending cycle of improvement with every new release.

This chapter was about the part of the Operator development cycle that is even more important than the first release: the next one. Releasing new software versions is not something that is exclusive to Operators, but the idiomatic processes and upstream Kubernetes standards do create a unique set of requirements for Operator projects. By exploring the technical steps necessary to create and publish a new version, alongside the more abstract policies and timelines that dictate guidelines for doing so, we have ensured that you are aware of a few suggestions for how to keep your Operator running for many releases to come.

This concludes the technical tutorial section of this book. While there are many topics and details that, unfortunately, did not fit within the scope of these chapters, we have explained all of the foundational concepts to build an Operator following the Operator Framework. In the next chapter, we'll summarize these concepts in an FAQ-style format to quickly refresh what was covered.

9
Diving into FAQs and Future Trends

The **Operator Framework** covers a lot of different topics, many of which have been discussed in this book. In this chapter, we will not discuss any new topics. Rather, we will revisit all of the main points that have been covered since the start of the book in short, digestible FAQ-style headings. It is the intent of these FAQs to provide a brief reference and refresher to everything that was covered during the course of this book. This should be a good, quick reminder of various topics in the event you are studying for an interview, certification exam, or just trying to remember an overview of a certain point that was made. The outline for these FAQs will fall under the following sections:

- FAQs about the Operator Framework

- FAQs about Operator design, **CustomResourceDefinitions** (**CRDs**), and APIs

- FAQs about the Operator SDK and coding controller logic

- FAQs about OperatorHub and the Operator Lifecycle Manager

- Future trends in the Operator Framework

These sections are roughly in the order that they appeared in the book, so reading this chapter in order will jog your memory and solidify your understanding of the topics as they were originally presented.

FAQs about the Operator Framework

These topics include an overview of the Operator Framework, its basic components, and the general vocabulary of Operator design. The topics from this section are from *Chapter 1, Introducing the Operator Framework*.

What is an Operator?

An Operator is a type of Kubernetes controller. Operators are designed to automate the management of Kubernetes applications and cluster components. They do this by continuously working to reconcile the current state of the cluster with the desired state, as defined by a user or administrator.

What benefit do Operators provide to a Kubernetes cluster?

Operators provide an idiomatic way for developers to encode automated cluster and application management logic into a controller. Operators also offer ways to expose the settings for this automation to non-developer users (for example, cluster administrators or customers). This automation frees up engineering and DevOps resources for many tasks.

How are Operators different from other Kubernetes controllers?

Operators are very similar to any other **Kubernetes** controller. Some examples of built-in controllers in Kubernetes include the **scheduler**, the API server, and the controller manager (which itself manages other controllers). These native controllers all handle the automated execution of core cluster functionality, such as placing Pods onto Nodes and maintaining Deployment replica counts. This is all part of the continuous state reconciliation pattern that Operators also exhibit.

However, while they are functionally similar, Operators are defined by conceptual and semantic conventions that differentiate them from standard controllers. These include the development libraries, tools, deployment methods, and distribution channels that comprise the Operator Framework.

What is the Operator Framework?

The Operator Framework is a set of development and deployment tools and patterns that define and support the standard processes for building an Operator. In broad terms, these include code libraries and scaffolding tools (the Operator SDK), a component for installing, running, and upgrading Operators (the **Operator Lifecycle Manager (OLM)**), and a centralized index for distributing Operators among the Kubernetes community (OperatorHub).

What is an Operand?

An Operand is the component or resources that are managed by an Operator.

What are the main components of the Operator Framework?

The main components of the Operator Framework are as follows:

- **The Operator SDK** – A set of common libraries and command-line tools for building an Operator from scratch. This includes wrappers for tools such as **Kubebuilder**, which are designed for generating code used in Kubernetes controllers.

- **The OLM** – A component designed to install, run, and upgrade (or downgrade) Operators in a Kubernetes cluster. Operator developers write (or generate) files that describe the Operator's relevant metadata in a way that the OLM can automate the Operator's deployment. The OLM also serves as an in-cluster catalog of installed Operators and can ensure that there are no conflicting APIs between installed Operators.

- **OperatorHub** – A centralized index of freely available Operators backed by an open source GitHub repository. Developers can submit Operators to `https://operatorhub.io/` for them to be indexed and searchable by users.

Now, let's talk about programming languages Operators can be written in.

What programming languages can Operators be written in?

Technically, an Operator can be written in any language that supports the necessary clients and API calls needed in order to interact with a Kubernetes cluster. But, the Operator SDK supports writing Operators in **Go** or building with tools such as **Helm** or **Ansible**. Helm and Ansible Operators are fairly straightforward to generate with the `operator-sdk` command-line tool, but these Operators are ultimately limited in their capability. In this book, we covered the code needed to write an Operator in Go, which offers far more functionality as defined by the Operator Capability Model.

What is the Operator Capability Model?

The Capability Model is a rubric for measuring the functionality an Operator provides and informing users of that functionality level. It defines five incremental levels of functionality:

1. **Level I – Basic Install**: Operators that are able to install an Operand, exposing configuration options for that installation if applicable

2. **Level II – Seamless Upgrades**: Operators that can upgrade themselves and their Operand without disrupting function

3. **Level III – Full Lifecycle**: Operators that can handle creation and/or restoration of Operand backups, failover scenarios for failure recovery, more complex configuration options, and scale Operands automatically

4. **Level IV – Deep Insights**: Operators that report metrics on themselves or their Operand

5. **Level V – Auto Pilot**: Operators that handle complex automated tasks including auto-scaling (creating more replicas of an Operand or deleting replicas, as needed), auto-healing (detecting and recovering from failure scenarios without intervention based on automated reporting such as metrics or alerts), auto-tuning (reallocating Operand Pods to better-suited Nodes), or abnormality detection (detecting when Operand performance does not align with usual application health)

These were some of the most fundamental topics covered in the first chapter. The next sections in this book built upon these to dive deeper into Operator design concepts.

FAQs about Operator design, CRDs, and APIs

These questions cover information about an Operator's design, including approaches to developing an Operator and how Operators can function within a Kubernetes cluster. The topics in this section were introduced in *Chapter 2, Understanding How Operators Interact with Kubernetes*, and *Chapter 3, Designing an Operator – CRD, API, and Target Reconciliation*.

How does an Operator interact with Kubernetes?

Operators interact with Kubernetes through event-triggered continuous monitoring of the cluster's state, wherein the Operator attempts to reconcile the current state with the desired state as specified by a user. From a technical standpoint, it does so through a standard set of Kubernetes client libraries that allow it to list, get, watch, create, and update Kubernetes resources.

What cluster resources does an Operator act on?

An Operator can act on any resource that is accessible through the Kubernetes API (and that the Operator has cluster permissions to access). This includes native Kubernetes resources (such as Pods, ReplicaSets, Deployments, Volumes, and Services) and **custom resources** (**CRs**) provided by third-party APIs or CRDs.

What is a CRD?

A CRD is a native Kubernetes API type that allows developers to extend the Kubernetes API with CR types that look and behave exactly like native Kubernetes API resources. Operator developers can create a CRD that defines their Operator's API type (for example, `customresourcedefinitions/MyOperator`) and this CRD then provides the template for creating CR objects that fit the definition of that type (for example, `MyOperators/foo`).

How is a CRD different from a CR object?

A CR object is the individual representation of an object that is based on the CRD template. In programming terms, it is the difference between an abstract type and an object instantiation of that type. CR objects are the API objects that users interact with to set Operator settings, for example with commands such as `kubectl get MyOperators/foo`.

What Kubernetes namespaces do Operators run within?

Operators can either be `namespaced` or `cluster-scoped`. Namespaced Operators run within an individual namespace, which allows multiple copies of the same Operator to be installed in a cluster. Cluster-scoped Operators run at the cluster-wide level, managing resources in multiple namespaces. The scope of an Operator is largely determined by the namespace scope defined in its CRD and the **role-based access control** (**RBAC**) policies assigned to the Operator's service.

How do users interact with an Operator?

Users first interact with an Operator by installing it. This can be either from an index such as OperatorHub or by installing directly from your organization's website or GitHub page. Operators can usually be installed with a single `kubectl create` command, especially when installed via the OLM.

Once installed, users will mainly interact with an Operator by creating a CR object that is a representation of its CRD. This CR object will expose API fields designed for tweaking the various settings associated with the Operator.

How can you plan for changes early in an Operator's lifecycle?

As with many software projects, thoughtful design allows for much easier growth as the project evolves. In the context of Operators, this means thinking early on about how the Operator (and more importantly, its Operand) may change over time. Upstream APIs and third-party dependencies may have support cycles that differ from your organization's own, so minimizing exposure to these dependencies can be very beneficial in reducing the work required later on. To this end, it can be helpful to start small with an Operator's design and build on that functionality as needed. This is part of the idea behind the Capability Model, with each level effectively building on the previous.

How does an Operator's API relate to its CRD?

The API that an Operator provides is the code definition of its CRD. When writing an API with the Operator SDK in Go, that API is generated into a CRD using the tools provided by the SDK.

What are the conventions for an Operator API?

Operator API conventions generally follow the same upstream Kubernetes conventions as for native API objects. The most important of these is that the Operator object contains two fields, `spec` and `status`, which provide the backbone for the cluster-state-reconciliation loop that Operators run on. `spec` is the user input section of the Operator object, while `status` reports the current functioning conditions of the Operator.

What is a structural CRD schema?

A structural CRD schema is an object definition that enforces known fields in cluster memory. Kubernetes requires CRDs to define a structural schema, which can be generated with tools provided by the Operator SDK. They provide security advantages and are often complex, owing to the recommendation that they are generated rather than hand-written.

What is OpenAPI v3 validation?

OpenAPI v3 validation is a format for providing field type and format validation upon the creation or modification of an object. Field validation is defined in the Go code for the CRD's API types. This validation is in the form of comments (for example, `//+kubebuilder:validation...`). These comments are generated into a validation schema when the Operator CRD is generated.

What is Kubebuilder?

Kubebuilder is an open source project that provides a tool for generating Kubernetes APIs and controllers. Many of the commands in the Operator SDK are wrappers for underlying Kubebuilder commands. This is good to know for debugging and support when troubleshooting Operator SDK issues.

What is a reconciliation loop?

The **reconciliation loop**, or **control loop**, is the main logic function of an Operator. It is conceptually the continuous cycle of checks an Operator performs to ensure that the actual cluster state matches the desired state. In reality, this is usually not done as a continuous loop but rather as an event-triggered function call.

What is the main function of an Operator's reconciliation loop?

The Operator's reconciliation loop is its core logic, during which the Operator evaluates the current state of the cluster, compares that to the desired state, and, if necessary, performs the required actions to reconcile the state of the cluster to match the desired state. This could mean updating a Deployment or tuning workload constraints to react to changing states.

What are the two kinds of event triggering?

Event triggering generally falls into two categories: **level-triggered** and **edge-triggered**. Operators are designed following the level-based triggering approach, in which a triggering event does not contain all of the context of the cluster state. Rather, the Operator must receive only enough information from the event to understand the relevant cluster state itself. By rebuilding this information each time, the Operator ensures that no state information is lost due to delays or dropped events. This is in contrast to edge-triggered events (reconciliation activated only by the incoming action of an event), which can result in information loss and are not suitable for large distributed systems such as Kubernetes. These terms stem from electronic circuit design.

What is a ClusterServiceVersion (CSV)?

A `ClusterServiceVersion` is a CRD provided by the Operator Framework that contains metadata describing a single version of an Operator. The `ClusterServiceVersion` CRD is provided by the OLM, which is the primary consumer of Operator CSVs. OperatorHub also uses the Operator CSV to present information to users about an Operator.

How can Operators handle upgrades and downgrades?

Operator versions are defined by their release version, image tag, and CSV metadata. The CSV in particular provides the concept of upgrade channels, which allow a developer to define subscription pathways for upgrade and downgrade versions. The OLM then knows how to transition installed Operators between versions thanks to this metadata. Operator API versions are defined in the Operator's CRD, which can contain information about multiple versions. This allows developers to ship an Operator release that supports multiple API versions simultaneously, enabling users to transition between versions within a single release.

How can Operators report failures?

Operators have several ways to report issues. These include standard runtime logs, metrics and telemetry, status conditions, and Kubernetes events.

What are status conditions?

Status conditions are an Operator's ability to use an upstream Kubernetes API type (v1. Condition) that quickly informs a user about various failure (or success) states via an Operator's status field in its CRD.

What are Kubernetes events?

Events are Kubernetes API objects that can be aggregated, monitored, and filtered by native Kubernetes tools such as kubectl. Their definition is richer than that of status conditions, allowing for more advanced information reporting to users.

FAQs about the Operator SDK and coding controller logic

The topics in this section focus on the technical development of an Operator with the Operator SDK. This includes generating an Operator project's initial boilerplate code, filling out the code with custom reconciliation logic, and expanding that code with more advanced features. These topics were introduced in *Chapter 4, Developing an Operator with the Operator SDK*, and *Chapter 5, Developing an Operator – Advanced Functionality*.

What is the Operator SDK?

The Operator SDK is a software development kit that provides code libraries and tools to quickly scaffold and build an Operator. It is mainly used through the operator-sdk binary, which provides commands to initialize projects, create boilerplate APIs and controllers, generate code, and build and deploy Operators in a cluster.

How can operator-sdk scaffold a boilerplate Operator project?

The first command to create an Operator SDK project is operator-sdk init. This command accepts additional flags to provide some project information (such as the code repository location of the project) that will later populate variables used when creating other aspects of the Operator (such as the API and controllers).

What does a boilerplate Operator project contain?

A boilerplate Operator project (that is, one that has just been created with `operator-sdk init` and no other changes) contains only a `main.go` file with some basic standard code, a `Dockerfile` file for building a container image, a `Makefile` file, which provides more commands to build and deploy an Operator, and some additional directories of config files and dependencies.

How can you create an API with operator-sdk?

The `operator-sdk create api` command initializes a template API file structure to be filled in by the developer. It accepts additional flags to define the API version and the Operator's resource name, and can even create the boilerplate controller for the Operator to handle the API object.

What does a basic Operator API created with operator-sdk look like?

The empty template API created by `operator-sdk create api` contains a basic definition for the Operator's `config` object. This object comprises upstream metadata types (containing fields such as `namespace` and `name`) as well as two sub-objects, representing the `spec` and `status` fields of the Operator's CRD.

What other code is generated by operator-sdk?

Along with the boilerplate template code, which is meant to be modified by the developer, the `operator-sdk` command also generates `deepcopy` and other code that is used by Kubernetes clients but should not be modified. Therefore, it's important to regularly re-run the code generators provided by `operator-sdk` to ensure this auto-generated code stays up to date.

What do Kubebuilder markers do?

Kubebuilder markers are specially formatted code comments placed on API object fields, types, and packages. They define field validation settings (such as type, length, and pattern) and other options that control the Operator's CRD generation. They allow such options to be configured in a location that makes them very clear, right next to the relevant code.

How does the Operator SDK generate Operator resource manifests?

The Operator SDK generates the relevant resource manifests (including the Operator's CRD, as well as other required resources such as **RoleBindings**) with the make manifests command. This command is defined in the standard Makefile file in a basic Operator project, and it invokes the controller-gen binary (part of the Kubebuilder toolset) to do the generation.

How else can you customize generated Operator manifests?

Generated Operator manifests can be customized beyond what is done by default with the controller-gen tool. This program is the underlying component responsible for much of the generated resource files, and running it manually offers access to additional commands and flags.

What are go-bindata and go:embed?

go-bindata and go:embed are two ways to compile raw files into Go code. The former is a package that can be imported into projects as a library, while the latter is a native Go compiler directive. Both are useful options for making Operator-related resources (such as an Operand Deployment YAML file) accessible in the code and readable for other users.

What is the basic structure of a control/reconciliation loop?

For most Operators, the basic structure of a control loop follows these steps:

1. Retrieve the desired configuration of the Operator/cluster.
2. Assess the current state of the cluster, with the information available to the Operator, and compare this to the desired configuration.
3. If needed, take action to transition the current cluster state toward the desired state.

At each of these steps, Operators should also implement error checking and status reporting branches to report any failures to the user (for example, if the Operator is unable to find its own configuration, it should obviously not proceed).

How does a control loop function access Operator config settings?

The main control loop function (`Reconcile()` in Operator SDK projects) accesses the Operator's config settings through its CR object. This requires that the user has created an instance of the CR object in the cluster (as defined by the CRD). Since CR objects are accessible through the Kubernetes API, the Operator is able to use Kubernetes clients and functions such as `Get()` to retrieve its config. In the Operator SDK, these clients are automatically populated and passed to the `Reconcile()` function, ready to use.

What information does a status condition report?

A `v1.Condition` object contains fields describing the status: `type` (a short name for the status), `status` (a Boolean value indicating the status), and `reason` (a longer description providing more information to the user). It also contains timestamp information regarding the last transition timestamp of the status.

What are the two basic kinds of metrics?

Metrics fall into roughly two categories: **service metrics** and **core metrics**. Service metrics are metrics that have been defined as custom for a specific component (or service), while core metrics are common metrics published by all services (such as CPU and memory usage).

How can metrics be collected?

Core metrics can be collected using the `metrics-server` component (`https://github.com/kubernetes-sigs/metrics-server`). Service metrics can be collected by any metrics aggregation tool, such as Prometheus.

What are RED metrics?

Rate, errors, and duration (**RED**) is an acronym describing the best practices for defining new metrics. It is recommended that the three key kinds of metrics for a service should include rate, errors, and duration. This means reporting the number of requests per period, the number of failed attempts, and the latency for a request, respectively.

What is leader election?

Leader election is the concept of running multiple copies of an application, with one copy (the leader) working at a time. This provides high availability for the application because if one replica fails, there are others ready to take its place. This concept applies to Operators, because it may be necessary to ensure Operator uptime in a distributed system.

What are the two main strategies for leader election?

Leader election can be implemented as either leader-with-lease or leader-for-life. The leader-with-lease approach is the default strategy for Operator SDK projects, and in it, the current leader makes periodic attempts to renew its status. This allows leader transitions to happen quickly, should they need to occur. In leader-for-life approaches, the leader only relinquishes its status when it is deleted. This makes recovery slower, but more definitive.

What are health and ready checks?

Health and ready checks are watchdog endpoints that allow an application to indicate when it is healthy (that is, running smoothly) and ready (that is, actively prepared to serve requests). The Operator SDK provides basic health and ready checks with a boilerplate project, but these can easily be extended to accommodate custom logic.

FAQs about OperatorHub and the OLM

These questions relate to the building, shipping, and deployment of Operators. Topics covered include installing Operators with the OLM and submitting Operators to OperatorHub. These topics come from *Chapter 6, Building and Deploying Your Operator*, and *Chapter 7, Installing and Running Operators with the Operator Lifecycle Manager*.

What are the different ways to compile an Operator?

Like many cloud-native applications, an Operator can be compiled either as a local binary or built into a container image suitable for deploying directly onto a Kubernetes cluster. The Operator SDK provides commands to do both.

How does a basic Operator SDK project build a container image?

The Operator SDK provides `Makefile` targets to build a Docker image with `make docker-build`. By default, this copies the main Operator source code (specifically, the main controller and API) along with its `assets` directory to the Docker image.

How can an Operator be deployed in a Kubernetes cluster?

Operators can be deployed manually once the Docker image has been built, similar to how any other application is deployed in a Kubernetes cluster. The Operator SDK simplifies this with the `make deploy` command. Operators can also be installed with the OLM.

What is the OLM?

The OLM is a component provided by the Operator Framework to manage the installation, running, and upgrade/downgrade of Operators in a cluster.

What benefit does running an Operator with the OLM provide?

The OLM provides a convenient tool to install Operators (as opposed to methods such as manually deploying the Operator image), as well as monitor the Operator status in a cluster. It can ensure that an Operator does not create any conflicts with other Operators. It can also handle upgrades and downgrades of Operators in the cluster. Finally, it makes the list of installed Operators in the cluster available to users.

How do you install the OLM in a cluster?

The OLM can be installed with the `operator-sdk` binary by running `operator-sdk olm install`.

What does the operator-sdk olm status command show?

Running `operator-sdk olm status` (after installing OLM in the cluster) shows the OLM's health by listing its required resources (including the CRDs it installs, its RoleBindings, Deployment, and namespaces).

What is an Operator bundle?

A bundle is a format for packaging an Operator's manifests and **ClusterServiceVersion (CSV)**.

How do you generate a bundle?

With the Operator SDK, a bundle can be generated by running `make bundle`. This command is interactive and will ask for information about the Operator and your organization, which will then be compiled into the Operator's metadata in the bundle.

What is a bundle image?

A bundle image is a container image that holds the information from the Operator's bundle. The image is used to deploy an Operator in a cluster, based on the underlying metadata.

How do you build a bundle image?

A bundle image can be built with the Operator SDK by running `make bundle-build`. This builds the basic Docker image that holds the bundle information.

How do you deploy a bundle with the OLM?

The `operator-sdk run bundle` command deploys a bundle image into a cluster with the OLM. This command takes an additional argument, which is the location of the bundle image in a container registry, for example, `operator-sdk run bundle docker.io/myreg/myoperator-bundle:v0.0.1`.

What is OperatorHub?

OperatorHub is a centralized index of published Operators from various developers and organizations. Its website is `https://operatorhub.io`. For each Operator, it provides information about the Operator and developer as well as installation guides and links to support resources and source code. It does this by parsing information from the Operator's bundle (mainly from the Operator CSV).

How do you install an Operator from OperatorHub?

Each Operator's page on OperatorHub includes a link to installation instructions. These instructions include a series of simple commands, usually using `kubectl create`:

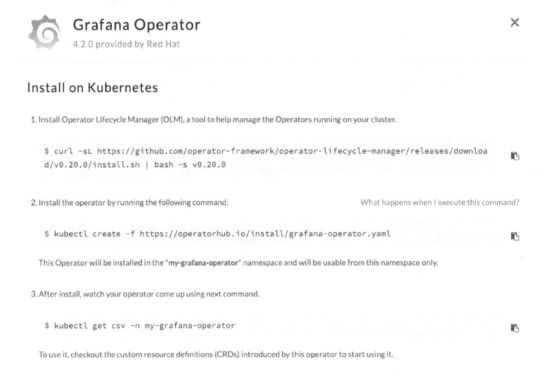

Figure 9.1 – Screenshot of OperatorHub installation instructions

How do you submit an Operator to OperatorHub?

Listing an Operator on OperatorHub involves creating a **pull request** (**PR**) to the GitHub repository, which serves as the backend index powering OperatorHub. The requirements include submitting a valid CSV file in a directory structure that is easily copied from the output of the Operator SDK bundle generation commands. Automated tests check that all requirements for submission have been met and merge the submission PR, with the Operator being listed on `https://operatorhub.io` soon after.

Future trends in the Operator Framework

This section refers to the future maintenance of your own Operator, as well as the alignment with ongoing work in the upstream Kubernetes community and how that relates to third-party Operator development. These topics are from *Chapter 8, Preparing for Ongoing Maintenance of Your Operator*.

How do you release a new version of an Operator?

Releasing a new version of an Operator is mostly dependent on your organization's release methods regarding timing and delivery infrastructure. But, the Operator Framework provides ways to denote your Operator's version in the form of the version field in the Operator's CSV (which is shown on OperatorHub), new API versions, and release channels.

When is it appropriate to add a new API version?

The most common time to add a new API version is when you are introducing a breaking change to the existing API. It can also be appropriate to increase the API's version as an indicator of its stability level (for example, promoting a `v1alpha1` API to `v1beta1` or `v1`). It is most important to follow the support timelines you have chosen to establish (or the Kubernetes timeline, if you have chosen to adopt those as a template) when replacing an old API version with a new one.

How do you add a new API version?

A new API can be added with the `operator-sdk create api` command. This command scaffolds the empty Go files just as when creating the Operator's initial API in a project. Once these files are filled out with the new API types, the corresponding generated code, CRD, and other manifests can be updated with the `make generate` and `make manifests` commands.

What is an API conversion?

An API conversion refers to an Operator's ability to transcribe API objects between incompatible versions of the same API. Conversion code in an Operator allows it to support multiple versions of its API within the same release. This is usually done by developers writing manual conversion logic to ensure that the incompatible fields between the two APIs can translate back and forth (round-trip) without losing any data. The biggest benefit of this is allowing users to transition from a deprecated API to a newer version seamlessly.

How do you convert between two versions of an API?

When designing a new API, consider how the existing information in an older version can be accurately translated to the new version and back. Then, you can convert between two versions of an API by implementing a conversion webhook in your Operator's code.

What is a conversion webhook?

A conversion webhook is an endpoint that is exposed in your Operator's Pod that accepts requests from the API server to encode and decode API objects between different versions.

How do you add a conversion webhook to an Operator?

A conversion webhook consists of two Go interfaces that must be implemented for the API object types that are being converted. These are the `Convertible` and `Hub` interfaces from `sigs.k8s.io/controller-runtime/pkg/conversion`. Most importantly, the `Convertible` interface requires two functions, `ConvertTo()` and `ConvertFrom()`, which are where a developer manually writes the logic to translate fields between the two objects. Then, the webhook manifests that expose the relevant endpoint can be created with the `operator-sdk create webhook` command. These manifests are then enabled by uncommenting the webhook references within the project's `Kustomization` files.

What is kube-storage-version-migrator?

kube-storage-version-migrator is a tool that allows Kubernetes users to manually convert their existing stored API objects to a new API version. This can be useful to help progress updates between Operator versions where a new API changes the storage version of the Operator's CRD object.

How do you update an Operator's CSV?

An Operator's CSV holds information about the Operator, including its current version. To increase the Operator's version, the CSV should first be updated to indicate the previous version of the Operator with the `replaces` field. In other words, this tells tools such as the OLM which version precedes the new version so that it can upgrade to the right version. Then, the `VERSION` variable in the Operator project's `Makefile` should be updated to the new version number (for example, `0.0.2`). The new CSV is generated with the `make bundle` and `make bundle-build` commands.

What are upgrade channels?

Upgrade channels are a way to offer different distribution upgrade pathways. For example, an Operator's CSV can define `alpha` and `stable` channels. Users can then choose to subscribe to either channel to get updated versions at the cadence and stability level they desire.

How do you publish a new version on OperatorHub?

OperatorHub hosts Operator information in bundles on GitHub. Each Operator has its own unique directory, with every different version of that Operator as its own subdirectory. These subdirectories each hold the bundle information (such as its CSV) for that version.

What is the Kubernetes deprecation policy?

The Kubernetes deprecation policy defines the API version support guidelines for core Kubernetes components. This provides downstream and third-party developers (as well as users) with a guaranteed timeline for support based on an API's stability. It is also a good template for other projects that are not strictly beholden to the policy but wish to align with the upstream Kubernetes policy for consistency.

How can API elements be removed in the Kubernetes deprecation policy?

In the Kubernetes deprecation policy, API elements can only be removed by incrementing the API version. In other words, if an Operator is following the Kubernetes deprecation policy, then it cannot remove a CRD field from the current API. It must, instead, create a new API version that does not have the field.

How long are API versions generally supported?

The support timeline for an API version in the Kubernetes deprecation policy depends on its stability level (alpha, beta, or GA). Alpha APIs have no guaranteed timeline and can be changed at any time. Beta APIs are supported for 9 months or three releases (whichever is longer). Graduated GA APIs cannot be removed, but they can be marked as deprecated.

How long is the Kubernetes release cycle?

The Kubernetes release cycle for a minor version is about 15 weeks long and contains several milestone dates, including Enhancements Freeze, Code Freeze, and Retrospective.

What is Enhancements Freeze?

Enhancements Freeze is the point during the release cycle at which **Kubernetes Enhancements (KEPs)** must be implementable and committed to the release or officially delayed to at least the next release.

What is Code Freeze?

At Code Freeze, all significant code changes must be merged into the current release or otherwise obtain an exception for a reasonable extension.

What is Retrospective?

Retrospective is a series of meetings leading up to and following the Kubernetes release in which community members reflect on the successful areas of the release and pinpoint processes that can use improvement.

How do Kubernetes community standards apply to Operator development?

Operator developers have no requirement to adhere to the Kubernetes community standards or participate in upstream development. However, an awareness of how the Kubernetes project functions is helpful to Operator developers because it offers guidelines for feature support, and Operators are very often dependent on certain Kubernetes features. These standards also provide a template for an Operator's own development processes that can be presented to the Operator's users.

Summary

There are obviously too many topics to cover in a discussion about the Operator Framework to distill them all into a single chapter of bite-sized trivia. But, to serve as a refresher (or crash course), the goal of this chapter was to recap the most important points that were already discussed, in a way that briefly explains them all, start to finish. For a deeper dive into any of these topics, the chapters they originated from have been provided as a resource. There are also any one of the many fantastic references that have been listed throughout this book to give support and further reading on all of these topics.

In the next chapters, we will examine real-world Operators that have been developed in the open source community. By doing so, we have the opportunity to relate each of the previous topics to a concrete example and compare the similarities and differences between what was done with the samples provided in this book and a real Operator.

10
Case Study for Optional Operators – the Prometheus Operator

The point of this book is to introduce, discuss, and demonstrate the main processes for developing an Operator for Kubernetes using the Operator Framework. To do this, a sample Operator with the basic functionality of managing an nginx deployment was built. This example was intended to serve as a tutorial on Operator development without overwhelming the reader with excessive features or the requirement of significant background knowledge to understand the use case. Hopefully, it has served that purpose well.

But the simplicity of that nginx Operator might make some of the steps in the Operator Framework seem excessive. It's also a big jump to go from learning about an example Operator to understanding the applications of real-world use cases. So, in this chapter, we will examine the Prometheus Operator (`https://prometheus-operator.dev/`), which is used to manage individual deployments of the Prometheus monitoring service (which was used to gather metrics from the nginx Operator earlier in the book). In this chapter, we are calling this an *optional* Operator because the Operand it manages is an application-level component and is not critical to the running of the cluster (in contrast, the next chapter will discuss how Operators can manage core cluster-level components). The Prometheus Operator will be discussed in the following sections:

- A real-world use case

- Operator design

- Operator distribution and development

- Updates and maintenance

Of course, while there are going to be many parallels to draw between the Prometheus Operator and the example nginx Operator from this book (which strictly followed the Operator Framework template), it is just as important to highlight the differences, too. Some of these will be covered throughout the chapter to show that even within the Operator Framework, there is no one-size-fits-all way to develop an Operator. That is the beauty of open source software such as this: its patterns and divergences promote a broad community of diverse projects.

A real-world use case

Prometheus (`https://github.com/prometheus/prometheus`) is a tool that is used for monitoring applications and clusters by collecting metrics exported by those applications and storing them in a time series manner. In *Chapter 5*, *Developing an Operator – Advanced Functionality*, we implemented basic Prometheus metrics in the nginx Operator to expose aggregate information about the total reconciliation attempts made by the Operator. This was just one small example of the potential application architecture designs that rely on Prometheus for monitoring.

Prometheus overview

Along with scraping and aggregating metrics, Prometheus also defines a data model for creating different types of metrics and implementing them in applications. This model is instrumented via the clients provided by Prometheus in various languages, including Ruby, Python, Java, and Go. These clients make it easy for application developers to export metrics in a format that is compatible with the Prometheus server's API (just as we did for the example nginx Operator).

Besides the counter metric type (which was used to sum the `reconciles_total` metric in the nginx Operator), the other metric types offered by Prometheus include Gauge, Histogram, and Summary. Each of these metrics can export additional attributes in the form of labels to give additional dimensions to the data they report.

In addition, Prometheus allows users to search metrics using its own query language called **PromQL**. The functionality of this language combined with the flexible and broad implementation possibilities of the metrics themselves has helped Prometheus grow to become one of the (if not the) leading metrics-gathering tools for cloud-native applications beyond just Kubernetes.

Earlier in the book, we briefly discussed how to use Prometheus clients to create new metrics and retrieve those metrics using PromQL (*Chapter 5, Developing an Operator – Advanced Functionality*) while also building the sample Operator. These topics, while important, do not relate much to the Prometheus Operator (regardless, they are briefly described here for the full context of the real-world use case). The more relevant aspects of Prometheus that the Operator addresses are the installation and configuration of Prometheus as an Operand.

Installing and running Prometheus

In *Chapter 6, Building and Deploying Your Operator*, we demonstrated one way to install Prometheus in a cluster by instrumenting the `kube-prometheus` library in the nginx Operator project. The advantage of kube-prometheus is that it installs a full monitoring stack, including components such as **Grafana**, but also including the Prometheus Operator itself. But what does it mean to install Prometheus in a cluster? And what steps do we save by using kube-prometheus (and, by extension, the Prometheus Operator)? To answer those questions, first, let's take a step back and understand how Prometheus works.

Central to an instance of Prometheus is the Prometheus server, which runs as a single binary that retrieves metrics and serves them to the web UI, notification services, or long-term storage. Similar to an Operator (or any application intended to be deployed to Kubernetes), this binary must be compiled and packaged into a container image. As described in the Prometheus documentation, the precompiled binary is available to download directly (as an executable or a container image), or it can be built from source (`https://github.com/prometheus/prometheus#install`). This is accessible enough for running locally, but for deployment to a Kubernetes cluster (especially a production one), further setup is required.

First, it is rarely acceptable to deploy a container directly into a cluster without some form of configuration. Kubernetes objects such as Deployments wrap the container in a managed and configurable representation that exposes options such as replica count and rollout strategies. So, installing Prometheus in a Kubernetes cluster manually would require defining the Kubernetes Deployment yourself.

Once it's running in a cluster, Prometheus then needs access to the applications that are exposing metrics. This requires additional resources such as `ClusterRoles` and `RoleBindings` to ensure the Prometheus Pod has permission to scrape metrics from the cluster and its applications. Those RBAC permissions must be bound to the Prometheus Pod via a `ServiceAccount` instance. Then, user access to the web UI requires a Service to make that frontend reachable in a web browser. That Service can only be exposed outside the cluster with an Ingress object.

These are already a lot of steps for an initial installation. However, managing that installation by hand also requires constant vigilance and a schematic understanding of each resource and its role. While certainly possible, having an Operator to handle these resources frees up engineering time and enables better scaling by abstracting complex manifest declarations.

However, as discussed throughout this book, many (if not most) Operators do more than simply install their Operand. Usually, they continue to manage the life cycle of the installed application, including allowing you to make changes to the running Operand's configuration. The Prometheus Operator does this for Prometheus, too.

Configuring Prometheus

As a full-featured application, Prometheus provides a rich set of configuration options to fit different scenarios. This configuration is documented in the official Prometheus documentation at `https://prometheus.io/docs/prometheus/latest/configuration/configuration/`. Within these settings, there are two sets of options for configuring Prometheus:

- Command-line flags: They control the settings that affect the Prometheus server itself, such as persistent storage access and logging settings.

- The YAML config: This is passed to Prometheus via a command-line flag (`--config.file` or `--web.config.file`) and provides declarative controls over the behavior of Prometheus' monitoring; for example, the metrics scraping settings.

This separation of setting types is a good design that is often employed in Kubernetes applications, Operators, and non-Kubernetes software. It has the benefit of clearly decoupling the runtime application settings from behavioral options, and this distinction is evident to users. However, from an administrative perspective, this creates two separate areas of concern that must be individually tracked.

Command-line flags

The full list of command-line flags that are available to the Prometheus binary is available by running `prometheus -h`. There are a few dozen options in total, but they are roughly organized into the following categories:

- Web

- Storage

- Rules

- Query

- Logging

Each of these categories has up to 10 (or more) individual settings controlling different aspects of the Prometheus server. In addition, there is the `--enable-feature` flag, which accepts a comma-separated list of features to enable (for example, `--enable-feature=agent,exemplar-storage,expand-internal-labels,memory-snapshot-on-shutdown` enables these four additional feature gates).

In a Kubernetes Deployment manifest, these flags would be controlled in the `spec.template.spec.containers.command` (or `.args`) field. For example, a simple Prometheus Deployment YAML with a config file passed to it and the preceding features enabled would look similar to the following:

```
apiVersion: apps/v1
kind: Deployment
metadata:
  name: prometheus
  labels:
    app: prometheus
spec:
  replicas: 1
  selector:
    matchLabels:
      app: prometheus
  template:
    metadata:
      labels:
        app: prometheus
    spec:
      containers:
      - name: prometheus
        image: docker.io/prom/prometheus:latest
        command: ["prometheus"]
        args:
        - --config=/etc/prom/config-file.yaml
        - --enable-feature=agent,exemplar-storage,expand-
internal-labels,memory-snapshot-on-shutdown
```

Of course, the config file also needs to be mounted into the Prometheus Pod so that it can be accessed, as shown in the following code. This shows the preceding Deployment YAML modified to add a `VolumeMount` instance, which makes the config file accessible to the Prometheus Pod as if it were a local file (the new code has been highlighted):

```
apiVersion: apps/v1
kind: Deployment
metadata:
```

```
   name: prometheus
   labels:
     app: prometheus
 spec:
   replicas: 1
   selector:
     matchLabels:
       app: prometheus
   template:
     metadata:
       labels:
         app: prometheus
     spec:
       containers:
       - name: prometheus
         image: docker.io/prom/prometheus:latest
         command: ["prometheus"]
         args:
         - --config=/etc/prom/config-file.yaml
         - --enable-feature=agent,exemplar-storage,expand-
internal-labels,memory-snapshot-on-shutdown
         volumeMounts:
         - name: prom-config
           mountPath: /etc/prom
       volumes:
       - name: prom-config
         configMap:
           name: prometheus-cfg
```

That config file (mounted as /etc/prom/config-file.yaml) will then need to be created as its own ConfigMap. This brings us to the second set of Prometheus options that the config file controls.

The YAML config settings

The Prometheus YAML configuration format exposes settings that control the general scraping (metrics-gathering) behavior of Prometheus. Among the available options are the platform-specific **Service Discovery** (**SD**) controls for individual cloud providers, including Azure, Amazon EC2, and Google Compute Engine instances. There are also options to relabel the metrics, enable the remote reading and writing of metrics data, and configure AlertManager notifications, tracing, and exemplars. Finally, the config offers TLS and OAuth settings for secure metrics scraping.

All of these options already present complex possibilities for a Prometheus config. Even the sample config file provided by Prometheus is almost 400 lines long! (However, it is intended to demonstrate many different types of metric setup. For example, take a look at `https://github.com/prometheus/prometheus/blob/release-2.34/config/testdata/conf.good.yml`.) This can quickly seem overwhelming, especially if you only want a simple metrics solution (as many users do). For this reason, we will mainly focus on a basic `scrape_config` section in a Prometheus config file. This is the main section of the config file that tells Prometheus where and how to find the metrics it is interested in.

This instruction is carried out by defining a series of `job` instances. Each job provides information about certain metrics targets and instructs Prometheus on how it can discover new metrics from targets that match those criteria. For example, the `kubernetes_sd_config` settings (which are relevant to scraping Kubernetes applications: `https://prometheus.io/docs/prometheus/latest/configuration/configuration/#kubernetes_sd_config`) can control metrics gathering for Nodes, Pods, Services, Endpoints, and Ingress objects.

Summarizing the problems with manual Prometheus

This chapter is not meant to be an introduction to how to run Prometheus. Rather, the intent of the earlier sections was to demonstrate, through specific examples, the potential complexities that can arise when running any sophisticated application and how these complexities multiply when that application is deployed to a platform such as Kubernetes, which demands its own maintenance overhead, too.

In summary, the problems discovered earlier fall into a few categories, as we will discuss next.

Excessive platform knowledge

From the very beginning (when installing Prometheus inside a cluster), it was necessary to know more about the platform and deployment resources than about running the actual application itself. From ClusterRoles and RoleBindings to even just the Deployment manifest declaration, an administrator must understand the Kubernetes installation architecture before they can even begin to run Prometheus itself.

This is bad because it distracts from engineering time, which could be otherwise allocated. However, it also creates an unstable environment, where this architectural knowledge is likely only learned once (at the time of installation) and promptly forgotten, or at the very least, not documented as well as other application-relevant resources. In the event of a disaster, this costs precious recovery time as the knowledge must be reacquired.

Complex configuration

Once Prometheus has been installed inside a cluster, immutable server settings must be passed via flags, and the individual metrics scraping jobs must be configured within a YAML file. For both of these steps, the vast number of available settings and flexible options for each setting present complex overall configurations. For metrics jobs, this complexity can potentially grow over time as more services are added to the cluster with metrics that must be gathered. This configuration must be maintained, and any changes need to be done with care to ensure they are rolled out effectively.

Restarts are required to enable changes

Speaking of changes, neither command-line flags nor config file settings take effect immediately. The Prometheus application must be restarted to notice the changes. This is not a big problem with changes to command-line flags, as doing so obviously requires the current running replica to stop (and, usually, making changes to a Kubernetes Deployment manifest will trigger a new replica with those changes anyway).

But it is less obvious for config file settings, which can lead to frustrating confusion as it might appear as though the changes are not taking effect. This might seem like a silly mistake, but it is one that is far too easy to make to consider risking it in a production environment. Even worse, it can lead to frustrated users making multiple changes at once before the mistake is realized, causing the new Deployment to pick up multiple unintended changes at once when it is finally restarted.

These are just a few simple examples of the problems that can be encountered when running applications without Operators. In the next section, we'll look more in detail at how the Prometheus Operator specifically approaches these issues with the goal of presenting an abstractable set of solutions that can be considered when building your own Operator for your application.

Operator design

The Prometheus Operator is designed to alleviate the issues mentioned earlier in regard to the complexity involved with running an instance of Prometheus in a Kubernetes cluster. It does so by abstracting the various configuration options that are available for Prometheus into **CustomResourceDefinitions** (**CRDs**), which are reconciled by the Operator's controllers to maintain that the cluster's Prometheus installation is consistent with the desired state, whatever that might be (and however it might change).

Of course, in contrast to the example nginx Operator from earlier tutorials, the Prometheus Operator manages a far more complex application with many more possible states that it must be able to reconcile. But the general approach is still the same, so we can evaluate this Operator through the lens of the same development steps that have been shown throughout this book.

CRDs and APIs

As discussed many times already, CRDs are the main objects upon which many Operators are built. This is because they allow developers to define custom API types that can be consumed by the Operator. Usually, this is how the user interacts with the Operator, setting their desired cluster state through the CRD that pertains to their Operator.

While this book has mainly focused on the concept of an Operator providing only a single configuration CRD (in the examples, this was just the `NginxOperators` object), the reality is that Operators can provide multiple different CRDs depending on their functionality. This is what the Prometheus Operator does. In fact, it provides eight different CRDs (which are described in detail at `https://github.com/prometheus-operator/prometheus-operator/blob/v0.55.1/Documentation/design.md`). The full list of available CRDs it provides defines the following object types:

- `Prometheus`
- `Alertmanager`
- `ThanosRuler`

- `ServiceMonitor`

- `PodMonitor`

- `Probe`

- `PrometheusRule`

- `AlertmanagerConfig`

We will discuss some of these object types in more detail next. In general, the purposes of these CRDs can be roughly broken down into a few categories:

- Operand deployment management

- Monitoring configuration settings

- Additional config objects

In order to keep the context of this chapter focused, we will only dive deeper into the first two groups of CRDs, as listed earlier. (The third, which is, here, referred to as *additional config objects*, includes the `Probe`, `PrometheusRule`, and `AlertmanagerConfig` types, which go into advanced monitoring settings that are beyond the scope of understanding Operator use cases.)

Operand deployment management

The first three CRDs, `Prometheus`, `Alertmanager`, and `ThanosRuler`, allow users to control the settings for Operand deployments. For comparison, our example `NginxOperator` CRD controlled the Kubernetes Deployment for an instance of nginx, exposing options such as `port` and `replicas`, which directly affected how that Deployment was configured. These Prometheus Operator CRDs serve the same purpose, just for three different types of Operand deployments. (Technically, the Prometheus Operator runs these Operands as StatefulSets, which is another type of Kubernetes object, not Deployments, but the same principles apply.)

These Operand-related CRDs are defined in the Operator's code at `pkg/apis/monitoring/v1/types.go` (note that the `pkg/api/<version>` pattern is similar to the one used in our Operator SDK code path). Talking specifically about the `Prometheus` object's top-level definition, it is exactly the same as our `NginxOperator` CRD:

prometheus-operator/pkg/apis/monitoring/v1/types.go:

```
type Prometheus struct {
    metav1.TypeMeta    `json:",inline"`
    metav1.ObjectMeta `json:"metadata,omitempty"`
    Spec PrometheusSpec `json:"spec"`
    Status *PrometheusStatus `json:"status,omitempty"`
}
```

With just the `TypeMeta`, `ObjectMeta`, `Spec`, and `Status` fields, this definition seems very straightforward. However, looking more closely at the `PrometheusSpec` object, the number of configuration options available becomes more evident:

prometheus-operator/pkg/apis/monitoring/v1/types.go:

```
type PrometheusSpec struct {
    CommonPrometheusFields `json:",inline"`
    Retention string `json:"retention,omitempty"`
    DisableCompaction bool
    WALCompression *bool
    Rules Rules
    PrometheusRulesExcludedFromEnforce []
PrometheusRuleExcludeConfig
    Query *QuerySpec
    RuleSelector *metav1.LabelSelector
    RuleNamespaceSelector *metav1.LabelSelector
    Alerting *AlertingSpec
    RemoteRead []RemoteReadSpec
    AdditionalAlertRelabelConfigs *v1.SecretKeySelector
    AdditionalAlertManagerConfigs *v1.SecretKeySelector
    Thanos *ThanosSpec
    QueryLogFile string
```

```
        AllowOverlappingBlocks bool
}
```

For the purposes of this chapter, it's not necessary to know what each option does. But the myriad of fields demonstrates how much an Operator's CRD can grow, emphasizing the need for careful management of an Operator's API. The list of available options goes even deeper with the embedded `CommonPrometheusFields` type, which offers controls over the number of replicas of Prometheus to run, the ServiceAccount settings for the Operand Pods, and a number of other settings related to the Prometheus deployment.

However, from a user's perspective, the `Prometheus` custom resource object they create in the cluster could look much simpler. This is because all of the fields in its type definition are marked with the `omitempty` JSON tag (for clarity, they were removed from all of the fields in the preceding code block except one). This denotes the fields as optional in the Kubebuilder CRD generator and does not print them if they are not set. Therefore, an example `Prometheus` object could be as simple as the following one:

```
apiVersion: monitoring.coreos.com/v1
kind: Prometheus
metadata:
  name: sample
spec:
  replicas: 2
```

Altogether, the `Prometheus` CRD offers a single spot for controlling some of the settings from either category, as discussed in the *Configuring Prometheus* section. That is, it exposes both command-line flag options and config file options in a single spot (along with Kubernetes-specific Deployment settings such as the replica count). It takes another step to disentangle some of these settings with the CRDs that control the monitoring options, which we will discuss next.

Monitoring configuration settings

While the `Prometheus` CRD allows users to configure the settings of the Prometheus metrics service itself, the `ServiceMonitor` and `PodMonitor` CRDs effectively translate to the Prometheus `job` config YAML settings, as described in the *Configuring Prometheus* section. In this section, we'll discuss how `ServiceMonitor` works to configure Prometheus to scrape metrics from specific Services (the same basic ideas apply to PodMonitor, which scrapes metrics from Pods directly).

To demonstrate this translation, the following `ServiceMonitor` object will be used to make the Prometheus Operator configure Prometheus so that it scrapes metrics from Service endpoints that are labeled with the `serviceLabel: webapp` labels:

```
apiVersion: monitoring.coreos.com/v1
kind: ServiceMonitor
metadata:
  name: web-service-monitor
  labels:
    app: web
spec:
  selector:
    matchLabels:
      serviceLabel: webapp
  endpoints:
  - port: http
```

More specifically, this object is broken down into two sections that are common to most Kubernetes objects: `metadata` and `spec`. Each serves an important role:

- The `metadata` field defines the labels that describe this `ServiceMonitor` object. These labels must be passed to the Prometheus Operator (in a `Prometheus` object, as described in the *Operand deployment management* section) to inform it about which `ServiceMonitor` objects the Operator should watch.

- The `spec` field defines a `selector` field, which specifies which application Services to scrape for metrics based on the labels on those Services. Here, Prometheus will ultimately know to scrape application metrics from Services labeled with `serviceLabel: webapp`. It will collect those metrics by querying the named `http` port on the Endpoints of each Service.

To gather this service discovery information (and, eventually, process it inside a Prometheus YAML configuration), the Prometheus Operator must be set up to watch `ServiceMonitors` with the `app: web` label. To do this, a `Prometheus` CRD object can be created similar to the following:

```
apiVersion: monitoring.coreos.com/v1
kind: Prometheus
metadata:
  name: prometheus
spec:
```

```
    serviceAccountName: prometheus
    serviceMonitorSelector:
      matchLabels:
        app: web
```

With this `Prometheus` object, the Prometheus Operator watches for instances of these `ServiceMonitor` objects and automatically generates the equivalent Prometheus YAML config. For the earlier `ServiceMonitor` object, that Prometheus configuration file looks similar to the following (note that this code is only shown here as an example to emphasize the complexity of a Prometheus config, and it is not necessary to understand it in depth):

```
global:
  evaluation_interval: 30s
  scrape_interval: 30s
  external_labels:
    prometheus: default/prometheus
    prometheus_replica: $(POD_NAME)
scrape_configs:
- job_name: serviceMonitor/default/web-service-monitor/0
  honor_labels: false
  kubernetes_sd_configs:
  - role: endpoints
    namespaces:
      names:
      - default
  relabel_configs:
  - source_labels:
    - job
    target_label: __tmp_prometheus_job_name
  - action: keep
    source_labels:
    - __meta_kubernetes_service_label_serviceLabel
    - __meta_kubernetes_service_labelpresent_serviceLabel
    regex: (webapp);true
  - action: keep
    source_labels:
    - __meta_kubernetes_endpoint_port_name
```

```
    regex: http
- source_labels:
  - __meta_kubernetes_endpoint_address_target_kind
  - __meta_kubernetes_endpoint_address_target_name
  separator: ;
  regex: Node;(.*)
  replacement: ${1}
  target_label: node
- source_labels:
  - __meta_kubernetes_endpoint_address_target_kind
  - __meta_kubernetes_endpoint_address_target_name
  separator: ;
  regex: Pod;(.*)
  replacement: ${1}
  target_label: pod
- source_labels:
  - __meta_kubernetes_namespace
  target_label: namespace
- source_labels:
  - __meta_kubernetes_service_name
  target_label: service
- source_labels:
  - __meta_kubernetes_pod_name
  target_label: pod
- source_labels:
  - __meta_kubernetes_pod_container_name
  target_label: container
- source_labels:
  - __meta_kubernetes_service_name
  target_label: job
  replacement: ${1}
- target_label: endpoint
  replacement: http
- source_labels:
  - __address__
  target_label: __tmp_hash
```

```
    modulus: 1
    action: hashmod
  - source_labels:
    - __tmp_hash
    regex: $(SHARD)
    action: keep
  metric_relabel_configs: []
```

Of course, this full YAML config is very long, and it would require significant effort to create (much less maintain) by hand. It's not important for the purpose of this discussion to explain the full config in detail. It is mainly shown here to emphasize the work done by an Operator to abstract such a complex configuration into a relatively simple CRD.

It is the relationship between CRDs such as `Prometheus` and `ServiceMonitor` that enables such abstraction in reasonable ways. For example, it would be easy to ship a single large `Prometheus` CRD that includes the settings for the monitoring services. This would also simplify the Operator's code by only requiring it to monitor one type of CRD for changes.

But decoupling these settings allows each CRD object to remain manageable and readable. Additionally, it provides granular access control over the modification of the Operand settings (in other words, specific teams can be granted the ability to create `ServiceMonitor` objects within their own namespaces in the cluster). This ad hoc configuration design gives cluster tenants control over the consumption of their own projects.

With this general understanding of the CRDs used by the Prometheus Operator (and how their design creates a cohesive API), next, we will look, in more detail, at how the Operator reconciles these objects from a technical perspective.

Reconciliation logic

To better understand the role of all the Prometheus Operator CRDs, it helps to know more about how the Operator configures the Prometheus Operand. Under the hood, the Prometheus Operator manages the Prometheus Pods' configuration through a secret (that is, the Kubernetes object designed to contain sensitive data of an arbitrary nature). That Secret is mounted into the Prometheus Pods as if it were a file and, thereby, passed to the Prometheus binary's `--config.file` flag.

The Operator knows to update this secret (and redeploy the Prometheus Operand, reloading the config file in the process) because it watches its `Prometheus` CRD (along with other CRDs such as `ServiceMonitor` and `PodMonitor`) in the cluster for changes.

Reloading Config Changes with Prometheus

Technically, the Prometheus Operator reloads the Prometheus config when it is changed without needing to redeploy the entire Operand. It does this with a sidecar container, running **prometheus-config-reloader** (`https://github.com/prometheus-operator/prometheus-operator/tree/main/cmd/prometheus-config-reloader`), which triggers a runtime reload by querying the `/-/reload` endpoint on the Prometheus server. While Prometheus supports runtime config reloading this way, many applications do not. So, for demonstration purposes, this chapter glosses over this technical detail to focus on the capabilities of the Operator and the more common use cases.

The Operator is able to monitor these CRD objects once it has been granted appropriate RBAC permissions to do so. This is because even though it defines those objects in its own project, the Kubernetes authentication services don't know that. To the cluster, the Operator is simply another Pod running an arbitrary application, so it needs permission to list, watch, get, or perform any other action on any type of API object.

In the nginx Operator example, the RBAC rules for accessing our Operator's CRD objects were automatically generated using Kubebuilder markers. Instead, the Prometheus Operator provides sample YAML files for its users with the appropriate permissions defined.

The Prometheus Operator creates three separate controllers for the three different Operands it supports (that is, Prometheus, Alertmanager, and Thanos). With the Operator SDK, the same design could be achieved by running `operator-sdk create api --controller` for each CRD that requires its own reconciliation logic.

Each controller watches for adds, updates, and deletes for the relevant objects that inform its reconciliation. For the Prometheus controller, these include the `Prometheus`, `ServiceMonitor`, and `PodMonitor` CRDs. But it also watches for changes to things such as Secrets and StatefulSets because, as mentioned earlier, these are the objects that are used to deploy the Prometheus Operand. So, by watching for these too, it can ensure that the Operand objects themselves are maintained at the appropriate settings and can recover from any deviations (for example, accidental manual changes to the Secret that holds the current Prometheus config YAML).

The main controller logic is implemented in a function called sync(), which is the equivalent to the Operator SDK's automatically created Reconcile() function. The sync() function follows the same general outline, as described for our sample nginx Operator, too. Some of the relevant code snippets from the Prometheus sync() function are detailed next.

First, the function gets the Prometheus CRD object that is necessary for the Prometheus Operand deployment to exist. If the object cannot be found, the controller returns an error. If it is found, then it creates a copy to work with:

```
func (c *Operator) sync(ctx context.Context, key string) error
{
    pobj, err := c.promInfs.Get(key)

    if apierrors.IsNotFound(err) {
        c.metrics.ForgetObject(key)
        return nil
    }
    if err != nil {
        return err
    }


    p := pobj.(*monitoringv1.Prometheus)
    p = p.DeepCopy()
```

Next, it parses the Prometheus object (and gathers any relevant ServiceMonitor objects or PodMonitor objects to parse too) in order to generate the YAML configuration Secret. This is done in a helper function that also checks for an existing Secret and creates one if none exist:

```
    if err := c.createOrUpdateConfigurationSecret(…); err !=
nil {
        return errors.Wrap(err, "creating config failed")
    }
```

Finally, it creates the Prometheus StatefulSet, which runs the Operand deployment. Similar to generating the config Secret, this part also uses helper functions to check for the presence of an existing StatefulSet and then decides whether to update it or create a new StatefulSet:

```
ssetClient := c.kclient.AppsV1().StatefulSets(p.Namespace)
...
obj, err := c.ssetInfs.Get(…)
  exists := !apierrors.IsNotFound(err)
  if err != nil && !apierrors.IsNotFound(err) {
    return errors.Wrap(err, "retrieving statefulset failed")
  }
...
sset, err := makeStatefulSet(ssetName)
  if err != nil {
    return errors.Wrap(err, "making statefulset failed")
  }
...
if !exists {
    level.Debug(logger).Log("msg", "no current statefulset
found")
    level.Debug(logger).Log("msg", "creating statefulset")
    if _, err := ssetClient.Create(ctx, sset, metav1.
CreateOptions{}); err != nil {
      return errors.Wrap(err, "creating statefulset failed")
    }
    return nil
}
...
level.Debug(logger).Log("msg", "updating current statefulset")
err = k8sutil.UpdateStatefulSet(ctx, ssetClient, sset)
```

This is equivalent to how the example nginx Operator created a Kubernetes Deployment object. However, rather than using a file-embedding library as we eventually did, the Prometheus Operator builds the StatefulSet object in memory. Without going into too much detail, that makes sense for this application because much of the StatefulSet definition is dependent on variable options that are set by logic in the code. So, there is not much advantage to maintaining an embedded file to represent this object.

Throughout this reconciliation loop, the Operator makes extensive use of structured logs and metrics to inform the user about its health. And while it doesn't report any `Condition` updates as our Nginx operator did, it does report other custom-defined fields in the `PrometheusStatus` field of the `Prometheus` CRD:

pkg/apis/monitoring/v1/types.go:

```go
type PrometheusStatus struct {
  Paused bool `json:"paused"`
  Replicas int32 `json:"replicas"`
  UpdatedReplicas int32 `json:"updatedReplicas"`
  AvailableReplicas int32 `json:"availableReplicas"`
  UnavailableReplicas int32 `json:"unavailableReplicas"`
}
```

This is a good demonstration of the fact that Operator CRDs can provide application-specific health information rather than only relying on existing patterns and upstream API types to convey a detailed status report. Combined with the fact that multiple `Prometheus` CRD objects can be created, with each representing a new deployment of Prometheus, the status information of individual Operand deployments is separated.

This is all a very high-level overview of the Prometheus Operator's reconciliation logic, with many specific implementation details omitted in order to draw more comparisons between the concepts discussed throughout the book related to Operator design.

Operator distribution and development

The Prometheus Operator is hosted on GitHub at `https://github.com/prometheus-operator/prometheus-operator`, where most of its documentation is also maintained. It is also distributed via OperatorHub at `https://operatorhub.io/operator/prometheus`.

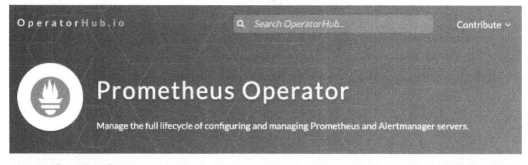

Figure 10.1 – The Prometheus Operator page on OperatorHub.io

As discussed in *Chapter 6, Building and Deploying Your Operator*, there are many different ways to run an Operator. From local builds to container deployments, each offers advantages for different development use cases. Then, in *Chapter 7, Installing and Running Operators with the Operator Lifecycle Manager*, the function of OperatorHub was explained as both a distribution index and an installation method when combined with the **Operator Lifecycle Manager (OLM)**.

In practice, this spectrum of distribution and installation options is illustrated by the Prometheus Operator. Inside its GitHub repository, the Prometheus Operator maintainers provide a single `bundle.yaml` file that allows curious users to quickly install all of the resources that are necessary to run the Operator with a simple `kubectl create` command.

Note that while this is similar in function to the bundle that's created to package an Operator for OperatorHub and OLM, technically, it is not the same. That's because it doesn't contain a **ClusterServiceVersion (CSV)** or other metadata that could be used by the OLM to manage an installation of the Prometheus Operator.

However, the Prometheus Operator does provide this information on OperatorHub. The backing CSV, along with the Operator's CRD files, are hosted for OperatorHub in its GitHub repository at `https://github.com/k8s-operatorhub/community-operators/tree/main/operators/prometheus`. This directory follows the same structure that was described in *Chapter 7, Installing and Running Operators with the Operator Lifecycle Manager*. Each new version of the Prometheus Operator's bundle is kept in its own numbered directory.

Figure 10.2 – The Prometheus Operator version directories on OperatorHub

The individual versions contain the YAML definitions for each CRD used by the Operator alongside a CSV that provides the metadata about the Operator and its resources.

To demonstrate the use case of this CSV, we can briefly look at some of the relevant sections, as shown in the following code. First, it describes the Operator's descriptive information, including its capability level (in this case, the Prometheus Operator is a Level IV Operator providing *Deep Insights* such as metrics about itself and its Operand):

prometheusoperator.0.47.0.clusterserviceversion.yaml:

```
apiVersion: operators.coreos.com/v1alpha1
kind: ClusterServiceVersion
metadata:
  annotations:
    capabilities: Deep Insights
    categories: Monitoring
    certified: "false"
    containerImage: quay.io/prometheus -operator/prometheus
-operator:v0.47.0
    createdAt: "2021-04-15T23:43:00Z"
    description: Manage the full lifecycle of configuring and
managing Prometheus and Alertmanager servers.
    Repository: https://github.com/prometheus -operator/
prometheus -operator
    support: Red Hat, Inc.
  name: prometheusoperator.0.47.0
  namespace: placeholder
```

Next, it embeds the various CRDs and a description of their fields. The following is an excerpt from the `Prometheus` CRD description:

```
spec:
  customresourcedefinitions:
    owned:
    - description: A running Prometheus instance
      displayName: Prometheus
      kind: Prometheus
      name: prometheuses.monitoring.coreos.com
      resources:
```

```
   - kind: StatefulSet
     version: v1beta2
   - kind: Pod
     version: v1
   - kind: ConfigMap
     version: v1
   - kind: Service
     version: v1
 specDescriptors:
 - description: Desired number of Pods for the cluster
   displayName: Size
   path: replicas
   x-descriptors:
   - urn:alm:descriptor:com.tectonic.ui:podCount
```

The CSV goes on to define the deployment of the Operator. This maps directly to the Kubernetes Deployment object, which will run the Operator Pods:

```
install:
  spec:
    deployments:
    - name: prometheus -operator
      spec:
        replicas: 1
        selector:
          matchLabels:
            k8s-app: prometheus -operator
        template:
          metadata:
            labels:
              k8s-app: prometheus -operator
          spec:
            containers:
            - args:
              - --prometheus-instance-
namespaces=$(NAMESPACES)
              - --prometheus-config-reloader=quay.io/
prometheus -operator/prometheus -config-reloader:v0.47.0
```

Finally, the CSV provides the RBAC permissions needed by the Operator to monitor its relevant resources in the cluster. Additionally, it also creates the RBAC permissions needed by the actual Prometheus Pods, which are separate from what the Operator needs. This is because the Operator and its Operand are separate entities in the cluster, and the Prometheus Server itself needs access to different resources to gather metrics (this is in contrast to the Prometheus Operator, which needs to access its CRDs).

Here are the RBAC permissions used to access its own CRDs, with wildcard (*) access under `verbs` indicating that the Operator can perform any API action against these objects (such as `get`, `create`, `delete`, and more):

```
permissions:
  - rules:
    - apiGroups:
      - monitoring.coreos.com
      resources:
      - alertmanagers
      - alertmanagers/finalizers
      - alertmanagerconfigs
      - prometheuses
      - prometheuses/finalizers
      - thanosrulers
      - thanosrulers/finalizers
      - servicemonitors
      - podmonitors
      - probes
      - prometheusrules
      verbs:
      - '*'
    serviceAccountName: prometheus -operator
```

The CSV concludes with contact information for the maintainers, along with links to the documentation and the version number of this release.

Offering a variety of distribution channels, in this case, GitHub and OperatorHub, has the obvious benefit of enabling an Operator to reach a broader audience of users. But this range of users can often be defined less by where the Operator is distributed and more by how the Operator is intended to be used. In other words, a user installing from OperatorHub is more likely to be evaluating (or actively running) the Operator in a production environment (with the full OLM stack) than a user installing the Operator from GitHub. In the latter case, such installations are probably more experimental, possibly from users seeking to contribute to the project themselves.

Accommodating the different use cases of an Operator in your distribution choices helps not only with the growth of a project but also its health. Recall that in *Chapter 2, Understanding How Operators Interact with Kubernetes*, we identified several types of potential users such as cluster users and administrators. While, in theory, an Operator's function might only apply to one type of user, the way that Operator is installed and run could vary for different kinds of users, including developers. Understanding these users and providing usage pathways for them increases the coverage of an Operator's functionality, improving the odds that bugs and potential features are identified.

As with many other topics in this book, these concepts are not specific to Operator design. But they are worth noting in the context of an Operator discussion to reiterate the ways in which they apply here. Similarly, while the topics of maintaining software and providing updates are not strictly specific to Operators, in the next section, we will still examine them through the lens of this Operator.

Updates and maintenance

The Prometheus Operator's community of maintainers is very active. With over 400 contributors to date (`https://github.com/prometheus-operator/prometheus-operator/graphs/contributors`), its code base remains fresh through ongoing maintenance. This allows the Prometheus Operator to publish regular releases on its GitHub Releases page (`https://github.com/prometheus-operator/prometheus-operator/releases`). As with any application, releasing regular updates promotes confidence in potential users by demonstrating an active investment in the project's support by its owners. In Kubernetes projects, such as Operators, this is even more important due to the relatively fast-paced and highly fluctuant developments in the underlying Kubernetes platform. Otherwise, with the current Kubernetes deprecation policy, an Operator might become unusable in new clusters in as little as a year (see *Chapter 8, Preparing for Ongoing Maintenance of Your Operator*).

In reality, in most cases, the main dependencies used by an Operator project will not frequently introduce breaking changes that require manual updates to remain compatible with existing users. Instead, most updates will simply be version bumps that bring in security, performance, and edge-case optimization improvements. To automate this process, the Prometheus Operator uses GitHub's Dependabot, which automatically creates pull requests to update any dependencies with new releases (`https://docs.github.com/en/code-security/dependabot`).

Automated dependency management tools such as Dependabot are a great way to ensure your Operator remains up to date with its dependencies, and thereby compatible with the most recent environment updates made by users. However, depending on your own scenario, you might still choose to manually update your Operator (for example, if you are aligning with a different release cadence where upstream patch releases might not be of significance to your own release).

Besides dependency updates, most Operators will also ship their own updates; for example, shipping a new API version (as covered in the *Releasing new versions of your operator* section of *Chapter 8*, *Preparing for Ongoing Maintenance*). In the case of the Prometheus Operator, the transition from API version `v1alpha1` to `v1` also involved a migration from Kubernetes **ThirdPartyResources** (a now-deprecated early implementation of extensible Kubernetes objects) to the at-the-time newly available CRDs (`https://github.com/prometheus-operator/prometheus-operator/pull/555`). So, examples of the Prometheus Operator shipping multiple API versions in a CRD are not currently available. However, as part of the project's roadmap, the intent is to update the `AlertManager` CRD from `v1alpha1` to `v1beta1`, leveraging conversion webhooks to translate between the two (this proposal is tracked and documented at `https://github.com/prometheus-operator/prometheus-operator/issues/4677`).

Finally, the Prometheus Operator maintains its own Slack channel for community support and discussion. Because the Operator is a third-party open source effort not directly affiliated with Prometheus itself, openly advertising the proper channels for support not only helps users find the right contacts but also respects the Prometheus maintainers' scope of responsibility. In this way, it is perfectly acceptable to publish an Operator that manages an Operand that you do not own. However, if this is not made clear, it can be very disruptive to your users and the owners of that Operand if the distinction between Operator and Operand is blurred.

Summary

In this chapter, we used the Prometheus Operator as an example to apply many of the concepts covered throughout the book. This Operator makes a good example because, aside from serving a common need by managing a popular application, it is actually one of the earliest Operators (having published its first release, v0.1.1, in December 2016). This predates the formalized Operator Framework, which developers can benefit from today, explaining idiosyncrasies such as its lack of Operator SDK libraries, but demonstrating the influence of early development decisions in the design of the Operator Framework.

At the beginning of this chapter, we gave a brief overview of Prometheus itself. This gave us a foundational understanding of the use case for a Prometheus Operator, particularly regarding the installation and configuration of Prometheus. This laid the groundwork to understand what the Prometheus Operator does to alleviate these pain points. By examining the CRDs it uses and how they are reconciled, we demonstrated how the Prometheus Operator abstracts that underlying functionality from the user, drawing parallels with the earlier chapters in the book (and the much simpler Nginx operator used to build the examples in those chapters). Finally, we looked at the more intangible aspects of the Prometheus Operator, such as its distribution and maintenance, to show how popular Operators apply these concepts from the Operator Framework.

In the next chapter, we will follow a similar case study but for a different Operator, that is, the etcd Operator.

11

Case Study for Core Operator – Etcd Operator

In the previous chapter (as for most of this book), we discussed Operators as a tool for managing applications that are deployed on Kubernetes. For most use cases, this is the main purpose of an Operator. In other words, the Operator serves to automate the applications that are developed by an organization. These applications are the products offered to users, and automating them helps to ship them without any issues and keep users happy. Beyond that, Kubernetes itself is simply the underlying architecture.

As a part of this, it's usually assumed that Kubernetes doesn't need any additional automation as would be provided by Operators. After all, it was a key point of the early chapters in this book that Operators are not functionally much different than the native suite of controllers that make up the Kubernetes control plane in the first place.

However, there are situations where an Operator can be used to manage aspects of core Kubernetes. While less common than application Operators, discussing some instances of core Operators helps to show the wide breadth of capabilities that the Operator Framework offers. Starting with a few of these examples, we will explore one in particular (though in less depth than the previous case study), the etcd Operator. Finally, we will explain some of the concepts around cluster stability and upgrades that are important to consider when developing Operators for Kubernetes. This will be done through the following sections:

- Core Operators – extending the Kubernetes platform
- etcd Operator design
- Stability and safety
- Upgrading Kubernetes

Core Operators – extending the Kubernetes platform

There is no differentiation in the Operator Framework between Operators that manage user-facing applications and infrastructure and Operators that manage core Kubernetes components. The only difference is simply in how the concepts of Operator design and development are applied to a slightly different class of problems. Still, the various Pods and control loops that comprise an installation of Kubernetes can be viewed as no different than the workload Pods that they deploy and manage.

Without getting too existential, this reduction bridges the conceptual gap between development *for* Kubernetes and the development *of* Kubernetes, making the latter seem much more approachable. This idea opens the gates to give system administrators and DevOps specialists greater control and flexibility over the cloud architectures they orchestrate.

Next, we will look at a few high-level examples of Operators that extend Kubernetes. We won't go into too much technical detail (such as their API or reconciliation logic), but we will briefly look at each of these examples to understand their use case and demonstrate some of the different ways that Operators can be used to directly manage Kubernetes system processes. The Operators we will look at are as follows:

- RBAC Manager (`https://github.com/FairwindsOps/rbac-manager`)
- Kube Scheduler Operator (`https://github.com/openshift/cluster-kube-scheduler-operator`)
- etcd Operator (`https://github.com/coreos/etcd-operator`)

Following this general overview, we will go into more technical detail about the etcd Operator in order to provide a similar understanding of the design concepts in this book, as we did in *Chapter 10, Case Study for Optional Operators – the Prometheus Operator*.

RBAC Manager

Role-Based Access Control (**RBAC**) policies are the cornerstone of Kubernetes authentication and authorization. RBAC settings in Kubernetes consist of three types of objects:

- **Roles** (or **ClusterRoles**, depending on the scope), which define the level of access that is allowed for a user or service
- **ServiceAccounts**, which are the identifying authorization object for a Pod
- **RoleBindings** (or **ClusterRoleBindings**), which map ServiceAccounts to Roles (or ClusterRoles)

These three types of Kubernetes API objects were explained in *Chapter 2, Understanding How Operators Interact with Kubernetes*. The relationship between them can be generally summarized by the following diagram (which was also used in that chapter):

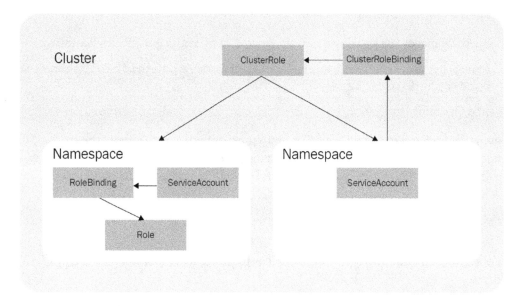

Figure 11.1 – A diagram of the RBAC object relationships

These objects allow for flexibility and control over the design of RBAC policies in a cluster. However, they can become confusing and cumbersome to manage, especially in large clusters with many different levels of access to different services. For example, if a user requires different authorization permissions for different namespaces, an administrator will need to create separate RoleBindings for each namespace with the appropriate grants. Then, if that user leaves the company or changes positions, the administrator will need to track each of those RoleBindings to ensure they can be appropriately updated. This approach is flexible, but it does not scale well for large organizations.

The **RBAC Manager** addresses these problems by providing a layer of abstraction on top of the native Kubernetes RBAC policy objects. This abstraction is represented by a single **CustomResourceDefinition (CRD)** that allows an administrator to effectively create and manage multiple RoleBindings for a user in one spot (with a slightly simplified syntax).

The effect of the RBAC Manager's simplified approach to authorization is that the management of RoleBindings is removed from a cluster administrator's manual responsibilities. This may be just one object in the chain of relational RBAC objects described previously, but it is the most repetitive and meticulous to track in large clusters. This is because the other objects, Roles/ClusterRoles and ServiceAccounts, will essentially map one-to-one against users, services, and access levels. But the intersection of users and access levels means that there is potentially a many-to-many relationship, held in place by RoleBindings.

The potential complexity of even a simple setup is shown by the following diagram, with four users each having varying levels of access (among hypothetical *read*, *write*, *view*, and *edit* roles). In this diagram, each arrow represents a RoleBinding that must be manually maintained:

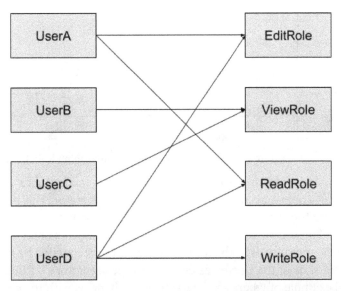

Figure 11.2 – A basic RoleBinding mapping of different users and RBAC levels

The RBAC Manager would simplify that same setup by inserting its CRD between the user and role definitions, creating a single access point to manage any user's permissions. Additionally, **UserB** and **UserC** can share an RBAC Manager CRD since they have the same role. This is shown in the following diagram:

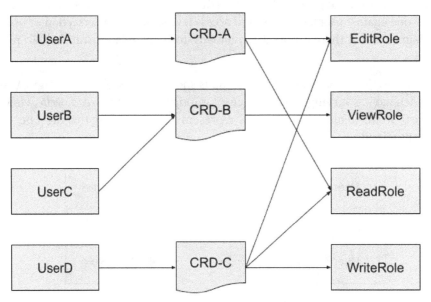

Figure 11.3 – A diagram of the RBAC Manager CRDs managing RoleBindings

In this setup, the individual arrows between CRDs and Roles (each still representing a single RoleBinding) are managed by the RBAC Manager Operator. This has the advantage of reducing the number of individual object relationships that administrators need to orchestrate. It also provides the state-reconciliation benefits of an Operator, wherein any updates or removals of the underlying roles are reconciled by the Operator to match the desired state of the cluster, as declared in the Operator's CRD objects. That behavior is a good example of where an Operator not only helps with the creation and management of complex systems but also ensures their ongoing stability.

The RBAC Manager is an Operator whose sole function is to manage native Kubernetes objects in the cluster. Next, we will discuss the Kube Scheduler Operator, which goes a step further to directly manage a critical component in the cluster, the Scheduler.

The Kube Scheduler Operator

The **Kube Scheduler** is one of the main control plane components in a Kubernetes cluster. It is responsible for assigning newly created Pods to Nodes, and it tries to do this in the most optimal way possible. This task is vital to the very function of Kubernetes as a cloud platform because if there is no way to schedule Pods onto Nodes, then the Pods cannot run their application code anywhere. And while manually deploying Pods onto specific nodes is possible, the automated evaluation and assignment done by the Scheduler obviously scales much better.

In addition, the definition of *optimal* Pod placement can be wildly different for different organizations (or sometimes just between different clusters from the same organization). For example, some administrators may want to spread the distribution of their workload Pods evenly among nodes to keep average resource consumption relatively low and prevent certain nodes from becoming overloaded. But other system admins may want the exact opposite, compacting as many Pods onto as few nodes as possible in order to minimize infrastructure costs and maximize efficiency. To accommodate these varying needs, the Scheduler provides a configuration API that allows you to customize its behavior.

The functionality and flexibility offered by the Scheduler are useful, but working with such an important part of a cluster can be risky. This is because if the Scheduler fails, then no other Pods can be scheduled (which includes some system Pods). Also, the complex configuration syntax for the Scheduler elevates the potential for this risk. For these reasons, many Kubernetes users shy away from Scheduler customization.

To address some of these issues, OpenShift (Red Hat's distribution of Kubernetes) ships with the **Kube Scheduler Operator** (in fact, OpenShift relies heavily on core Operators, which is discussed more thoroughly at `https://www.redhat.com/en/blog/why-operators-are-essential-kubernetes`). This Operator is built using an Operator library developed specifically for OpenShift Operators rather than the Operator SDK. This allows the Kube Scheduler Operator to manage the health and stability of the critical Scheduler Pods in a way that is consistent with the other features built into OpenShift. While most Operator developers will not need to write their own development libraries, this example shows that in certain use cases, it's fine to do so if you have unique needs that the Operator SDK does not support.

The Kube Scheduler Operator does follow other design aspects of the Operator Framework, such as the use of CRDs as the primary interface between users and Operator logic. This Operator makes use of two CRDs. One is used to configure Operator-specific settings and report the health status of the Operator through Conditions, while the other holds the Scheduler configuration options that control how the Scheduler assigns Pods to nodes. The Operator goes a step further with the second CRD by predefining sets of Scheduler configurations for common use cases, completely abstracting the underlying Operand settings into easily understood pick-and-choose options.

The role of the Kube Scheduler Operator as a system Operator, managing a core component of Kubernetes clusters, is an important task. Its function serves the critical purpose of placing Pods onto appropriate nodes, and its ability to recover from failures helps maintain cluster health. In the next section, we will look at one more Operator that performs similar management of another critical component, etcd.

The etcd Operator

etcd (https://etcd.io/) is the primary key-value store that backs Kubernetes clusters. It is the default option for persistent storage of the API objects that exist in a cluster, preferred for its scalable and distributed design that makes it optimal for high-performance cloud computing.

The **etcd Operator** is designed to manage the etcd component in a cluster. Even though it is no longer actively maintained, its GitHub repository is still available in an archived state to provide a historical reference for future developers. For the purpose of this chapter, the preserved status of the etcd Operator offers a permanent, unchanging reference for the design of a core Operator.

etcd clusters within a Kubernetes cluster can be managed by the etcd Operator in a full variety of functions. These include creating etcd instances, resizing federated installations of etcd, recovering from failures, upgrading etcd without suffering uptime, and performing backups of etcd instances (as well as restoring from those backups). This suite of functionality qualifies the etcd Operator as a Level III Operator in the Capability Model. If you recall from *Chapter 1*, *Introducing the Operator Framework*, Level III Operators are referred to as **Full Lifecycle** Operators, indicating their ability to manage Operands beyond simple installation and support advanced management operations, such as upgrades and backups.

Installing and managing etcd manually in a Kubernetes cluster is a fairly advanced task for most users. The majority of Kubernetes developers take the availability of a persistent data store for granted, assuming that all of their objects and cluster state information will always be available. But if the etcd processes fail, there is the potential for it to have catastrophic effects on the entire Kubernetes cluster.

Similar to any other database, the etcd component in a cluster is responsible for storing all objects that exist in the cluster. Failure to do so can bring even basic cluster functionality to a halt. Such a failure could be caused by a bug, an incompatible API, or even by malformed input when trying to modify the etcd installation (for example, scaling it to provide higher availability). Therefore, a smooth-running cluster is dependent on efficient and accurate access to data in etcd.

The etcd Operator aims to simplify the management of etcd by automating the operational commands required to create, resize, upgrade, back up, and recover etcd clusters through the Operator's various CRDs. In the next section, we will go into more detail about the CRDs that the Operator uses to do this and how those CRDs are reconciled to ensure that the current state of etcd in the cluster matches the administrator's desired state.

etcd Operator design

Like most other Operators, the etcd Operator is built with CRDs as its focal interface for user interaction. Understanding the CRDs an Operator provides is a good way to get a basic understanding of how the Operator works, so that is where we will begin our examination of the etcd Operator.

CRDs

The three CRDs used by the etcd Operator are EtcdCluster, EtcdBackup, and EtcdRestore. The first CRD, EtcdCluster, controls the basic settings for the etcd installation, such as the number of Operand replicas to deploy and the version of etcd that should be installed. A sample object based on this CRD looks like the following:

simple-etcd-cr.yaml:

```
apiVersion: etcd.database.coreos.com/v1beta2
kind: EtcdCluster
metadata:
  name: example
spec:
  size: 3
  version: 3.5.3
```

In this example, were this object to be created in a cluster (`kubectl create -f simple-etcd-cr.yaml`), it would instruct the etcd Operator to create three replicas of etcd version 3.5.3. Besides these options, the `EtcdCluster` CRD also provides configuration settings for specifying a specific repository to pull the etcd container image from, Operand Pod settings (such as `affinity` and `nodeSelector` settings), and TLS config.

The other two aforementioned CRDs, `EtcdBackup` and `EtcdRestore`, work in tandem to allow users to declaratively trigger the backup and subsequent restoration of etcd data in a cluster. For example, etcd can be backed up to a **Google Cloud Storage** (**GCS**) bucket by creating the following custom resource object:

etcd-gcs-backup.yaml:

```
apiVersion: etcd.database.coreos.com/v1beta2
kind: EtcdBackup
metadata:
  name: backup-etcd-to-gcs
spec:
  etcdEndpoints:
    - https://etcd-cluster-client:2379
  storageType: GCS
  gcs:
    path: gcsbucket/etcd
    gcpSecret: <gcp-secret>
```

This instructs the Operator to back up the etcd data available at the cluster endpoint `https://etcd-cluster-client:2379` and send it to the GCS bucket called `gcsbucket/etcd`, authenticated by the **Google Cloud Platform** (**GCP**) account secret pasted in `<gcp-secret>`. That data can be restored by later creating the following `EtcdRestore` object:

etcd-gcs-restore.yaml:

```
apiVersion: etcd.database.coreos.com/v1beta2
kind: EtcdRestore
metadata:
  name: restore-etcd-from-gcs
spec:
  etcdCluster:
```

```
    name: sample-etcd-cluster
  backupStorageType: GCS
  gcs:
    path: gcsbucket/etcd
    gcpSecret: <gcp-secret>
```

These CRDs make it much easier to perform backups of etcd data and restore those backups by abstracting and automating the heavy lifting into Operator controller logic, but they also ensure that these operations are performed successfully. By eliminating the need for human interaction for the bulk of the backup, they also eliminate the possibility of human error in collecting and transferring the etcd data to a backup location. If the Operator does encounter an error when performing the backup, this information is reported through the `Status` section of the custom resource object for that backup.

Reconciliation logic

In the etcd Operator's design, each CRD type has its own reconciliation controller. This is good Operator design, as recommended by the best practices in the Operator Framework's documentation. Similar to the Prometheus Operator from *Chapter 10, Case Study for Optional Operators – the Prometheus Operator*, each controller monitors for cluster events, involving the CRD it reconciles. These events then trigger a level-based reconciliation loop that either creates (or modifies) an etcd cluster, performs a backup of a running etcd cluster, or restores the already backed-up data from an earlier etcd cluster.

Part of this reconciliation logic involves reporting the status of the Operand etcd Pods. After receiving an event and during the ensuing reconciliation cycle, if the Operator detects a change in the etcd cluster's status, it reports on this through the matching custom resource object's `Status` field. This type has the following sub-fields:

```
type ClusterStatus struct {
    // Phase is the cluster running phase
    Phase   ClusterPhase `json:"phase"`
    Reason string        `json:"reason,omitempty"`


    // ControlPaused indicates the operator pauses the control
of the cluster.
    ControlPaused bool `json:"controlPaused,omitempty"`


    // Condition keeps track of all cluster conditions, if
they exist.
```

```
    Conditions []ClusterCondition
`json:"conditions,omitempty"`


    // Size is the current size of the cluster
    Size int `json:"size"`


    // ServiceName is the LB service for accessing etcd nodes.
    ServiceName string `json:"serviceName,omitempty"`


    // ClientPort is the port for etcd client to access.
    // It's the same on client LB service and etcd nodes.
    ClientPort int `json:"clientPort,omitempty"`


    // Members are the etcd members in the cluster
    Members MembersStatus `json:"members"`
    // CurrentVersion is the current cluster version
    CurrentVersion string `json:"currentVersion"`
    // TargetVersion is the version the cluster upgrading to.
    // If the cluster is not upgrading, TargetVersion is
empty.
    TargetVersion string `json:"targetVersion"`
}
```

Note that the etcd Operator implements its own `ClusterCondition` type (rather than the `Condition` type used in *Chapter 5, Developing an Operator – Advanced Functionality*). This is because active maintenance of the etcd Operator was archived shortly before the native `Condition` type was merged into upstream Kubernetes. However, we mention this here as another tangible example where awareness of upstream **Kubernetes Enhancement Proposals** (**KEPs**) and their status throughout the timeline of a release cycle can have an impact on third-party Operator development (see *Chapter 8, Preparing for Ongoing Maintenance of Your Operator*).

Beyond just reconciling the desired state of the etcd Operands and reporting their status, the etcd Operator also has reconciliation logic to recover from failure states.

Failure recovery

The etcd Operator collects status information on the etcd Operands and, based on the availability of a majority of the Operand Pods (known as a quorum), attempts to recover the etcd cluster if necessary. In the edge case where an etcd cluster has no running Operand Pods, the Operator must make an opinionated decision whether to interpret this as an etcd cluster that has completely failed or simply one that has yet to be initialized. In this case, the Operator always chooses to interpret it as a failed cluster and attempts to restore the cluster from a backup, if one is available. This is not only the simplest option for the Operator but also the safest (see the following *Stability and safety* section), as the alternative (assuming no availability is simply uninitialized) could result in ignored failure states that would otherwise be recovered from.

Besides recovering from failures in its Operands, the etcd Operator also has reconciliation logic to recover from failures in itself. Part of this recovery path hinges on the etcd Operator registering its own CRD in the cluster. This is interesting because it contradicts the best practices recommended by the Operator SDK documentation. But by coupling the creation of the CRD with the Operator (rather than having an administrator create it – for example, with `kubectl create -f`), the Operator can use the presence of that CRD to determine whether it exists as a new installation or has previously been running in a cluster.

This is applicable because when any Pod restarts, including Operator Pods, they will not have any inherent knowledge about their predecessors. From that Pod's perspective, it has begun life with a fresh start. So, if the etcd Operator starts and finds its CRD already registered in a Kubernetes cluster, that CRD serves as a sort of canary indicator to inform the Operator that it should begin reconstructing the state of any existing Operand Pods.

Failure recovery is one of the most helpful benefits that Operators provide because the automated error handling allows small hiccups in the day-to-day operation of a cluster to be quickly and gracefully absorbed. In the case of the etcd Operator, it makes opinionated decisions in handling failures and managing its own CRD to create a support contract that clearly defines its recovery procedures. Doing so contributes greatly to the cluster's overall stability, which is the focus of the next section.

Stability and safety

Applications and components in a Kubernetes cluster can occasionally be prone to unexpected failures, such as network timeouts or code panics due to unforeseen bugs. It is part of an Operator's job to monitor for these spontaneous failures and attempt to recover from them. But of course, human error in the pursuit of adjusting the system can be another source of failure. As a result, any interaction with or modification of the core system components in Kubernetes brings inherent risk. This is elevated because manual adjustments to one component can contain errors (even minor ones) that cause a domino effect, as other components that depend on it begin reacting to the original error.

Perhaps the prime objective of an Operator is to provide stability and safety in production environments. Here, stability refers to the ongoing performant operation of the Operand programs, and safety is the ability of an Operator to sanitize and validate any inputs or modifications to that program. Think of an Operator like a car, whose purpose is to run its motor smoothly along the road while allowing the driver to control it within reasonable parameters. In Kubernetes, Operators for system components offer some of the leading examples of designs where exceptional care has been taken to offer a safe design that lends itself to stable functioning.

The Kube Scheduler Operator from our earlier example ensures cluster stability by taking an opinionated approach toward the available configuration options it provides. In other words, the total number of possible scheduling settings is restricted (via predefined collections of options) to only those arrangements that have been tested and known to have minimal risk to the cluster. Then, changes to these settings are rolled out in a predetermined manner by the automation code in the Operator. The combination of not only automating changes but restricting the available changes significantly reduces any opportunity for users to make an error when updating their cluster. This abstraction is shown in the following diagram, where the user only needs to interact with the CRD, while all fields in the predefined settings are hidden away as black-box options:

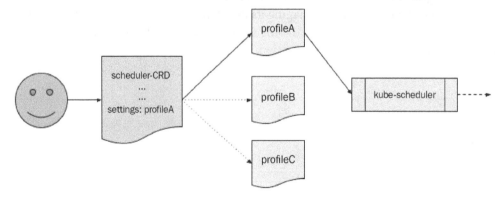

Figure 11.4 – A diagram of the Kube Scheduler Operator config abstraction

The etcd Operator has similar safeguards in place. For example, uninstalling the Operator does not automatically remove the custom resource objects associated with it. While some Operators may implement this sort of garbage collection as a feature to make uninstallation easier, the developers of the etcd Operator intentionally chose not to do so in order to prevent accidental deletion of running etcd clusters.

Another approach the etcd Operator takes toward achieving operational stability is in the spreading of its Operand Pods. As mentioned earlier, the etcd Operator allows users to configure individual Pod settings for the etcd Operand instances it deploys, such as `nodeSelectors`. One interesting field to note, however, is that it provides a simple Boolean option called `antiAffinity`. This value can be set within the `EtcdCluster` CRD, as follows:

```
spec:
  size: 3
  pod:
    antiAffinity: true
```

Enabling this value serves as shorthand to label the Operand etcd Pods with an `antiAffinity` field to themselves. The effect of this is that the individual Pods will not be scheduled onto the same nodes. This has the benefit of ensuring that the Pods are distributed in a highly available fashion so that if one node goes down, it does not risk bringing down a significant number of the etcd Pods.

The effective change of this one line on the Operand Pods looks something like the following:

```
apiVersion: v1
kind: Pod
metadata:
  name: etcd-cluster-pod
spec:
  affinity:
    podAntiAffinity:
      requiredDuringSchedulingIgnoredDuringExecution:
      - labelSelector:
          matchLabels:
            etcd_cluster: <etcd-resource-name>
        topologyKey: kubernetes.io/hostname
```

Packaging this into a single line not only saves time but also saves users from potential mistakes. For example, the following snippet looks very similar to the preceding one. However, it has the exact opposite effect, requiring that all similar etcd Pods are scheduled onto the same node:

```
apiVersion: v1
kind: Pod
metadata:
  name: etcd-cluster-pod
spec:
  affinity:
    podAffinity:
      requiredDuringSchedulingIgnoredDuringExecution:
      - labelSelector:
          matchLabels:
            etcd_cluster: <etcd-resource-name>
        topologyKey: kubernetes.io/hostname
```

If this mistake went unnoticed and the node that hosted all of the etcd Pods went down, it would result in potentially the entire cluster being unable to function.

While the `antiAffinity` setting was eventually deprecated in favor of allowing users to provide full Pod Affinity blocks (as shown previously), its presence offers an example of the safe configuration that Operators can provide. Swapping that with the option of full Affinity blocks has the trade-off of safety for flexibility, which is a delicate scale that Operator developers must balance in their own implementations.

These are just a few examples of the ways in which Operators can package complex operations into abstracted user-facing settings to provide a safe interface and stable cluster management. Next, we'll look at how Operators can maintain that stability during a potentially tumultuous period of upgrades.

Upgrading Kubernetes

In software, upgrades are usually a fact of life. Very rarely is a single piece of software that lives for more than a brief time able to run continuously without upgrading its code to a new version. In *Chapter 8, Preparing for Ongoing Maintenance of Your Operator*, we discussed ways to prepare and publish new version releases of an Operator. In that chapter, we also explained the different phases of the Kubernetes release cycle and how they impact Operator development. This is especially true for Operators that are deeply entrenched in the Kubernetes platform components.

For system Operators, fluctuating changes in the upstream Kubernetes code base are often more than just simple features. For example, when `kube-scheduler` was refactored to accept an entirely different format of configuration (referred to as the Scheduler Framework, which is no relation to the Operator Framework – see `https://kubernetes.io/docs/concepts/scheduling-eviction/scheduling-framework/` for more details – though they are not technically relevant here), the predefined settings profiles handled by the Kube Scheduler Operator's CRD needed to be completely rewritten to accommodate that change. System Operators absolutely must be aware of changes such as this between Kubernetes versions because in many cases, they themselves are performing the upgrade to a new version.

The etcd Operator has logic to handle upgrades of its Operand, etcd. This adds a point of complexity, as etcd is technically not part of the Kubernetes payload that runs a cluster. So, while the upstream Kubernetes developers must coordinate their releases to support certain versions of etcd (as etcd developers work on their own releases), the etcd Operator developers need to react to changes in both dependencies.

In this way, the Operand is also a dependency for an Operator, which comes with its own constraints that must be respected. For example, when performing an upgrade, the etcd Operator first verifies that the desired new version is allowed by the upgrade strategies supported by etcd. This specifies that etcd can only be upgraded by at most one minor version at a time (for example, from 3.4 to 3.5). If more than this is specified, the etcd Operator can determine that the upgrade should not proceed and aborts the process.

If an upgrade is allowed to proceed, the etcd Operator updates each of its Operand Pods with a **rolling upgrade** strategy. This is a common way to upgrade versions of applications, and even Kubernetes itself, where one Pod (or Node, in the case of Kubernetes) in the available set is upgraded at a time. This allows for minimal downtime, but it also allows for the ability to roll back (revert to the previous working version) if an error is encountered when upgrading any single instance. This is a critical stability benefit for any Operator to support as it provides the opportunity to diagnose the error in a stable environment. The etcd Operator uses its own backup and restoration workflows to handle these kinds of rollbacks during an upgrade.

Upgrading Kubernetes (as with upgrading any software) can be a tedious process, which is why it's not uncommon for cluster administrators to delay upgrading until the last possible minute. Unfortunately, this only exacerbates the problem, as more incompatible changes are introduced with each skipped version. But introducing Operators to help with handling these changes can make upgrades much smoother or, at the very least, assist with collecting information to diagnose failed upgrades. While most Operator developers will not be writing Operators for system-level Kubernetes components, the lessons from those who have will serve as examples of how to handle upgrades in even the most mission-critical scenarios.

Summary

In this chapter, we took one final look at some examples of Operators responsible for some of the most critical tasks in a Kubernetes cluster. These core, or system, Operators automatically manage the most complex and delicate workflows for Kubernetes administrators while balancing functionality and care for the importance of their Operands. From these types of Operators, we can draw lessons about the fullest spectrum of capabilities in the Operator Framework.

The intent of this chapter wasn't to offer these examples as tutorials or imply that many developers will write their own Operators to manage core Kubernetes components. Rather, they serve as extreme cases where concepts from the Operator Framework were applied to outlier problem sets. But understanding the edge cases in any problem is the best way to form a strong understanding of the entire problem.

We did this by beginning the chapter with a brief overview of three different system Operators. Then, we took a deeper dive into the technical details behind the etcd Operator, understanding how it uses CRDs and reconciliation logic to manage the data backup storage for Kubernetes. Finally, we concluded by exploring how Operators such as the etcd Operator provide a stable and safe interface for their tasks, even when upgrading versions of Kubernetes.

With that, we end this introduction to the Operator Framework. The topic of Kubernetes Operators is a broad one, with many potential technical details to explore and volumes of excellent literature already written by well-qualified authors. It would be difficult to succinctly compile all of the available information on Operators from every available resource, so in this book, we simply tried to cover the most important topics in novel and interesting ways. Hopefully, you found this book insightful and helpful in building your own understanding of the Operator Framework and will find ways to apply these lessons when building your own Operators. Happy hacking!

Index

A

Abnormality Detection 18
advanced functionality
  need for 102, 103
Ansible 9, 226
Apache 2.0 License 178
API conversions
  implementing 194-198
API directory
  generating 189, 190
API elements
  removing, in Kubernetes
    deprecation policy 241
API version
  adding 239
  adding, need for 239
  adding, to Operator 187-189
  converting 240
  deploying 203-205
  support timeline 241
  testing 203-205
application programming interface (API)
  about 4, 72-76, 254, 255
  designing 48
  overview 227

applications
  managing with, Operators 19
Auto-Healing Operators 18
Auto Scaling Operators 18
Auto-Tuning Operators 18
AWS 173

B

BinData 81, 82
boilerplate Operator project
  config files and dependencies,
    containing 232
  operator-sdk init command,
    creating 231
bundle
  about 12, 236
  generating 237
bundle/ directory
  manifests/ 166
  metadata/ 166
  tests/ 166
bundle image
  about 163, 237
  building 237
  deploying, with OLM 237

C

Call for Exceptions 217
Capability Model
  about 8, 14, 226
  Level I-basic install 15
  Level II-seamless upgrades 15
  Level III-full lifecycle 16
  Level IV-deep insights 17
  Level V-auto pilot 18
  Operator functions, defining with 14-18
  reference link 14
cert-manager
  reference link 203
ci.yaml
  reference link 179
cluster administrators 32, 33
ClusterRoleBindings 277
ClusterRoles 76, 277
cluster-scoped 31
ClusterServiceVersion (CSV) 61, 230, 236
clusters, without Operators
  cluster states change, reacting to 5, 6
  managing 4
  sample application 4, 5
cluster users 33, 34
code
  generating, by
    operator-sdk command 232
Code Freeze 215, 218, 242
Code Thaw 220
coding controller logic
  overview 231
command-line flags 249-251
community Operators repository
  reference link 208

container image
  building 131, 132
  Operator, building locally 133
  Operator image, building with
    Docker 133, 135, 136
continuous integration (CI) 57
controller-gen 79
control loop function
  about 229
  Operator config settings, accessing 234
  structure 233
conversion webhook
  about 240
  adding, to Operator 240
core metrics
  about 114, 234
  collecting, with
    metrics-server component 234
core Operators
  about 276
  etcd Operator 282, 283
  Kube Scheduler Operator 281, 282
  RBAC Manager 278-280
CRD schema 51, 52
CRD versioning
  reference link 187
CR object
  about 227
  versus CustomResourceDefinition
    (CRD) 227
Custom Resource (CR) 7, 48
CustomResourceDefinition (CRD)
  about 103, 27, 254, 278
  configuration settings,
  monitoring 257-261
  designing 48

Operator API, relating to 228
overview 227
object types 254
operand deployment
    management 255, 257
updating 192, 193
overview 227
versus CR object 227
custom service metric
adding 115

D

deployments 27
Deprecated API Migration Guide
    reference link 214
deprecation and backward compatibility
    planning 209-211
development-operations (DevOps) 6
disaster recovery (DR) 16
Docker
    Operator image, building with 134-136
    troubleshooting 149-151
    URL 149
Dockerfile syntax
    reference link 149
DockerHub
    URL 137
downgrades
    handling 60-62

E

edge-based triggering 58
edge-triggered 230
end users and customers 35
Enhancements Freeze 215-217, 242

errors
    reporting, with events 64, 65
    reporting, with logging 62, 63
    reporting, with status updates 63, 64
etcd Operator
    about 282, 283
    reference link 277, 282
etcd Operator, design
    about 283
    CRDs 283, 284
    failure recovery 287
    reconciliation logic 285, 286
event triggering
    about 230
    edge-triggered 230
    level-triggered 230

F

failure reporting
    using 62
Full Lifecycle Operators 282

G

go:embed marker
    about 233
    using, to access resources 82, 83
garbage collection (GC) 63
GA release 219
generated Operator manifests
    customizing 233
Go 9, 70
go-bindata access resources
    old version 84
    using 82, 83
Google Cloud 173

Google Cloud Platform
    (GCP) account 284
Google Cloud Storage (GCS) bucket 284
Grafana
  about 247
  reference link 114
Grafana Operator 174

H

health and ready checks 235
health checks
  adding 125, 126
Helm 9, 226

J

JavaScript Object Notation (JSON) 56

K

kind
  reference link 148
  troubleshooting 148
  URL 137
kind codebase
  reference link 148
Kubebuilder 53, 76, 225, 229
Kubebuilder markers 232
Kubebuilder Metrics
  reference link 115
Kubebuilder resources 99
KubeCon 214
Kube Controller Manager 7
Kubelet 27
kube-prometheus
  about 247
  configuring 144, 145

installing 144
reference link 144
Kubernetes
  about 4, 103
  issues 98
  Operator, interacting with 227
  upgrading 290, 291
Kubernetes API design conventions
  following 49, 50
Kubernetes cluster
  benefits, to Operators 224
  Operator, deploying 236
Kubernetes cluster resources
  CustomResourceDefinition
      (CRDs) 28, 29
  deployments 27
  interacting with 26
  namespace 31
  Pods 27, 28
  ReplicaSets 27, 28
  RoleBindings 29, 30
  roles 29, 30
  ServiceAccounts 29, 30
Kubernetes community
  repository, reference link 221
  working with 220, 221
Kubernetes community standards
  applying, to Operator development 242
Kubernetes controller
  versus Operator 224
Kubernetes deprecation policy
  API elements, removing 241
  reference link 211
Kubernetes Enhancement
    Proposal (KEP) 216
Kubernetes events 231

Kubernetes objects
  metadata field 258
  spec field 258
Kubernetes platform
  extending 276
  release, overview 214-216
Kubernetes Release Cadence Change
  reference link 214
Kubernetes release cycle
  about 242
  Call for Exceptions 217
  Code Freeze 218
  Enhancements Freeze 216, 217
  GA release/Code Thaw 219, 220
  Retrospective 220
  start 216
  Test Freeze 218
Kubernetes release timeline
  aligning with 214
Kubernetes resources 97, 98
Kubernetes SD configurations
  reference link 252
Kubernetes Slack server
  URL 148
kube-scheduler 114
Kube Scheduler Operator
  about 281, 282
  reference link 277
Kubernetes standards, complying
    with for changes
  about 211, 212
  API conversion 213
  API lifetime 213
  APIs, removing 212
kube-storage-version-migrator
  about 240
  reference link 205

Kustomize
  URL 146

L

leader 121
leader election
  about 235
  implementing 121-125
  strategies 235
leader-for-life 122
leader-with-lease 122
level-based event triggering
  versus edge-based event triggering 58
level-triggered 230

M

maintainers
  identifying 32, 36
Make
  reference link 148
Makefile issues
  troubleshooting 148
metrics
  about 234
  core metrics 234
  service metrics 234
  troubleshooting 152, 153
metrics reporting
  custom service metric, adding 115
  implementing 114, 115
  RED metrics 116-118, 120
metrics-server application 114
metrics-server component
  about 234
  used, for collecting core metrics 234

Microsoft 173
minimum viable product (MVP) 48, 210

N

namespace 31 159
namespace-scoped 31
nginx Pod
    managing, with simple Opertaor 48-66
no-operation (no-op) function 54, 74, 126

O

OLM Operator condition
    using 111-114
OLM support
    troubleshooting 181
OpenAPI version 3 (V3)
        validation 53, 229
OpenTelemetry
    reference link 114
Operand 7, 225
Operator
    about 69, 224
    applications, managing with 19
    API 72-76
    API version, adding to 187-189
    beneficial features, designing 36, 37
    benefit, providing to
        Kubernetes cluster 224
    building locally 133
    bundle files, exploring 166-168
    bundle image, building 169-171
    bundle image, pushing 171
    changes, pushing 143
    changes, testing 143
    cluster service version (CSV) file 165
    compiling 235

control loop, writing 89-97
conversion webhook, adding to 240
CRD conditions 104-110
CustomResourceDefinition
    (CRD), updating 192, 193
deploying, in Kubernetes cluster 236
deploying, in test cluster 137-143
downgrades, handling 230
errors, troubleshooting 97
failures, reporting 231
implementing, on cluster resources 227
installing, from OperatorHub 238
interacting, with Kubernetes 227
Kubernetes namespaces,
        running within 228
new versions, releasing of 187
Operator bundle, deploying
        with OLM 172
Operator bundle, generating 163-165
programming languages, writing 226
project, setting up 71, 72
redeploying, with metrics 146
resource manifests, adding 76-81
running 163
safety 288-290
stability 288-290
submitting, to OperatorHub 238
troubleshooting 148
upgrades, handling 230
users, interacting with 228
version, releasing 239
versus Kubernetes controller 224
Operator API
    conventions 229
    relating, to CRDs 228
Operator bundle 163, 236
Operator capabilities
    applying 20-22

Operator Capability Model
  about 226
  functionality level 226
Operator changes
  deprecating 40
  designing 38
  iterating 39
  planning 38
Operator config settings
  accessing, with control
    loop function 234
Operator CRD
  example 53, 54
Operator CSV version
  updating 206, 207
Operator design
  overview 227
Operator development
  Kubernetes community standards,
    applying to 242
Operator, errors troubleshooting
  Kubebuilder resources 99
  Kubernetes resources 97, 98
  SDK resources 99
Operator for Prometheus
  reference link 17
Operator Framework
  about 3, 6, 225
  components 225
  Custom Resource 7
  future maintenance 239
  Kubernetes controllers 6, 7
  Operand 7
  overview 8, 9, 224
  summarizing 19, 20
Operator functions
  defining, with Capability Model 14-18

OperatorHub
  about 187, 237, 225
  documentation, reference link 177
  new version, releasing on 208, 209
  Operators, installing from 174-177, 238
  Operator, submitting 178-80
  overview 235
  URL 173, 225
  version, publishing 241
  working with 173
OperatorHub.io
  about 12
  Operators, distributing 12, 13
  overview 12
OperatorHub repository 179
OperatorHub support
  troubleshooting 181-183
Operator image
  building, with Docker 133-136
Operator lifecycle
  changes, planning 228
Operator Lifecycle Manager (OLM)
  about 61, 103, 159, 225, 236
  benefit 236
  installing, in cluster 236
  installing, in Kubernetes cluster 159-161
  Operators, managing with 11
  overview 235
  used, for deploying bundle image 237
  working with 161-163
Operator's Cluster Service Version (CSV)
  about 13
  updating 240
Operator SDK
  about 69, 187, 225, 231
  developing 9-11
  overview 231

resources 99
used, for generating resource
manifests 233
operator-sdk api command
about 232
creating 232
operator-sdk command
used, for generating code 232
Operator SDK documentation 122
operator-sdk init command
creating, in boilerplate
Operator project 231
operator-sdk olm status command
displaying 236
Operator SDK project
container image, building 235
Operators, significance in Kubernetes
reference link 281
operator upgrades, in Operator
Lifecycle Manager (OLM)
reference link 206

P

Pod 27, 159
project manifests
updating, to deploy webhook 199-203
Prometheus
about 24
community, reference link 151
configuring 249
installing 247, 248
issues, summarizing with 252-254
overview 247
real-world use case 246
reference link 114, 118
running 247, 248
URL 151

prometheus-config-reloader 262
Prometheus configuration
command-line flags 249-251
reference link 249
YAML config 249, 252
Prometheus Operator
APIs 254
CRDs 254, 255
designing 254
reconciliation logic 261-265
reference link 144
used, for reloading config changes 262
PromQL 247
pull request (PR) 178, 238

Q

quorum 287

R

rate, errors, and duration
(RED) metrics 234
RBAC Manager
about 278-280
reference link 277
RBAC settings
RoleBindings 277
Roles 277
ServiceAccounts 277
reconciliation loop
about 229
function 230
structure 233
Red Hat 173
RED metrics
about 116-120
duration 116

Errors 116
Rate 116
release candidate (RC) 215
release phases, Kubernetes
reference link 216
ReplicaSets 27
required resources
working with 55-57
resource embedding
simplifying 87-89
resource manifests
about 81, 82
adding 76-81
generating, with Operator SDK 233
Retrospective 216, 220, 242
Role-Based Access Control (RBAC)
about 80
policies 277
RoleBindings 159, 233, 277
Roles 277
rolling upgrade strategy 291

S

scheduler 224
Scheduling Framework
reference link 291
ServiceAccounts 159, 277
Service Discovery (SD) controls 252
service metrics 114, 234
ServiceMonitor
reference link 146
split-brain scenario
about 122
solving, in leader-with-lease 123
status condition report
information 234

status conditions
about 231
reporting 103
storage version 192
structural CRD
schema 229
structural schema 52

T

target reconciliation loop
designing 57-60
edge-based, versus
level-event triggering 58
test cluster
Operator, deploying in 137-143
Test Freeze 218
troubleshooting
additional errors 153
Docker 149-151
kind 148
Makefile issues 148
metrics 151, 152
OLM support 181
OperatorHub support 181-183

U

upgrade channels 241
upgrades
handling 60-62
user groups
cluster administrators 32
cluster users 32
end-users/customers 32
user interface (UI) 54

users
  identifying 32
  interacting, with Operator 228

V

versions
  releasing, of Operator 187
  releasing, of OperatorHub 208, 209

W

webhook
  deploying, by updating project
    manifests 199-203

Y

YAML Ain't Markup Language
    (YAML) 10
YAML config
  about 249
  settings 252

Packt.com

Subscribe to our online digital library for full access to over 7,000 books and videos, as well as industry leading tools to help you plan your personal development and advance your career. For more information, please visit our website.

Why subscribe?

- Spend less time learning and more time coding with practical eBooks and Videos from over 4,000 industry professionals

- Improve your learning with Skill Plans built especially for you

- Get a free eBook or video every month

- Fully searchable for easy access to vital information

- Copy and paste, print, and bookmark content

Did you know that Packt offers eBook versions of every book published, with PDF and ePub files available? You can upgrade to the eBook version at packt.com and as a print book customer, you are entitled to a discount on the eBook copy. Get in touch with us at customercare@packtpub.com for more details.

At www.packt.com, you can also read a collection of free technical articles, sign up for a range of free newsletters, and receive exclusive discounts and offers on Packt books and eBooks.

Other Books You May Enjoy

If you enjoyed this book, you may be interested in these other books by Packt:

The Kubernetes Bible

Nassim Kebbani, Piotr Tylenda, Russ McKendrick

ISBN: 9781838827694

- Manage containerized applications with Kubernetes
- Understand Kubernetes architecture and the responsibilities of each component
- Set up Kubernetes on Amazon Elastic Kubernetes Service, Google Kubernetes Engine, and Microsoft Azure Kubernetes Service
- Deploy cloud applications such as Prometheus and Elasticsearch using Helm charts
- Discover advanced techniques for Pod scheduling and auto-scaling the cluster
- Understand possible approaches to traffic routing in Kubernetes

Learning DevOps - Second Edition

Mikael Krief

ISBN: 9781801818964

- Understand the basics of infrastructure as code patterns and practices

- Get an overview of Git command and Git flow

- Install and write Packer, Terraform, and Ansible code for provisioning and configuring cloud infrastructure based on Azure examples

- Use Vagrant to create a local development environment

- Containerize applications with Docker and Kubernetes

- Apply DevSecOps for testing compliance and securing DevOps infrastructure

- Build DevOps CI/CD pipelines with Jenkins, Azure Pipelines, and GitLab CI

- Explore blue-green deployment and DevOps practices for open sources projects

Packt is searching for authors like you

If you're interested in becoming an author for Packt, please visit authors.
packtpub.com and apply today. We have worked with thousands of developers and
tech professionals, just like you, to help them share their insight with the global tech
community. You can make a general application, apply for a specific hot topic that we are
recruiting an author for, or submit your own idea.

Share Your Thoughts

Now you've finished *The Kubernetes Operator Framework Book*, we'd love to hear your
thoughts! Scan the QR code below to go straight to the Amazon review page for this book
and share your feedback or leave a review on the site that you purchased it from.

https://packt.link/r/1803232854

Your review is important to us and the tech community and will help us make sure we're
delivering excellent quality content.